The Laws of the Sun

The Laws of the Sun

The Spiritual Laws & History Governing Past, Present & Future

Ryuho Okawa

Lantern Books · New York
A Division of Booklight Inc.

Lantern Books
One Union Square West, Suite 201
New York, NY 10003

Library of Congress Cataloging-in-Publication Data

Okawa, Ryuho, 1956–
    [Taiyo no ho. English]
    The laws of the sun: the spiritual laws & history governing past,
present & future / Ryuho Okawa.
        p. cm.
Translated by The Institute for Research in Human Happiness Ltd.
ISBN 1-930051-62-X (alk. paper)
1. Kofuku no Kagaku (Organization)—Doctrines. I. Kofuku no
Kagaku (Organization) II. Title.

BP605.K55 T3513 2001
291.4—dc21
                                                    2001037740

printed on 100% post-consumer waste paper, chlorine-free

Table of Contents

Chapter Three: The River of Love

Chapter Four: The Ultimate Enlightenment

Preface

This edition of *The Laws of the Sun* was published in 1994 in Japan. Since that time, the book has sold more than seven million copies, making my name known throughout Japan. There has been constant international press coverage, both about our organization, the Institute for Research in Human Happiness (IRH), and me. This included a full page in the London *Financial Times*, and a major article in the US *Wall Street Journal.*

Although this is one of more than three hundred books I have published, it is a vital work that represents the core of my thoughts. Not only does it reveal secrets of multi-dimensional worlds, of the creation of the universe, and the truth of all human history, but it also gives precise directions for individuals in pursuit of enlightenment and training for spiritual awareness. As its content shows, this is the genesis of our time, the scripture of the new age. This new thinking, which integrates Christianity and Buddhism, is the new way of salvation, and it will form the new humanity of the coming age.

The nature of this book is infinitely mystical. Do not try to rely upon your "common sense" to comprehend it. Rather, I challenge you to transform your common sense with its contents. My sincere wish is that people who accept the content of this book as the new common sense will become new leaders

amid the political and economic chaos of the world, and will
be able to guide humanity toward true happiness.

Ryuho Okawa
President
The Institute for Research in Human Happiness

One: When the Sun of Truth Rises

1. The Sun of Buddha's Truth

"Supreme Consciousness," "Eternal Law"—both of these are descriptions of the Truth of Buddha. Buddha's Truth is a golden thread woven through the past, present, and future of the lives of every person who has ever lived and ever will live. It sustains all of the sacred traditions of the world including, for instance, Buddhism, Confucianism, and the Gospel of Love preached by Jesus Christ.

But the eternal truths that salve the heart of humanity number more than merely those first woven into the weft some two or three thousand years ago. In Japan alone, after all, Kukai, and others spread Buddhism during the Heian period (AD 794–1185); the Kamakura period that followed (1185–1333) saw Honen, Shinran, Eisai, Dogen, Myoe, Nichiren, and Ippen bring about a Buddhist revival, while Rennyo effected the restoration of the Buddhist sect called the True Pure Land School. At other times, Nagarjuna lived and philosophized in India as did T'ien-t'ai Chih-i in China. All these represent dif-

ferent strands within the tapestry woven by the golden thread
of Buddha's Truth.

We need all the eternal truths that we can get in a world in
which materialism seems to be running out of control and out-
pacing spiritual maturity. It is time to feel the lifegiving
warmth of the Sun of Buddha's Truth. Just as the sun in the sky
supplies the energy of physical life on Earth, the Sun of
Buddha's Truth supplies humankind continuously with inner
life force. Just as the sun is always shining even when it is
obscured by clouds from the people on the ground, there are
times when the Buddha Sun is hidden from our sight, when
people may become confused and start to believe that it has
gone forever, yet it is only temporarily obscured and always
reappears. Returning to the world after a period of darkness, it
may be greeted as "the light of salvation," "the light of
redemption," or even "the light of life."

It is this Truth that I want to pass on in this book, *The Laws
of the Sun*, particularly to those for whom the Sun of Truth
seems to have disappeared below the horizon. A powerful light
is emanating from my own small corner of the planet.
Certainly, the modern world needs light—it needs a veritable
Sun of Truth, rising to illuminate the world and banish the
darkness and illusion that we are evidently so good at surrounding ourselves with.

Buddha's Truth is a lifegiving message, and it is in the
earnest hope of passing on that message that this book is written, in the hope that many people who read it or come into

contact with it will take its message as the goal of their lives. Behind every single word I have written lies my fervent desire to bring salvation to the world; each word is imbued with life and light, and I would dearly love to think that all the words together will be treasured by my brothers and sisters through out the world.

2. What Is Buddha?

Many of us have at one time or another given some thought to the true meaning of life. But what conclusions did we reach? In thinking about life, the first step is to decide exactly what the word "life" means. To most people it is the period of just a few decades that we spend in this world, between our birth and our death—and if this is the way you think, then your whole outlook on life should undergo a fundamental change once you have read this book. If this is all life is, it would mean that a person born at a specific time and given a specific name by his or her parents, leaves nothing behind at death but a name, a few bones, and a handful of ash. Everything else is converted into carbon dioxide or water vapor and dissipates in the air.

If this is the case, what on earth do we spend our lives working for? What is the use of studying? Why do we put ourselves to such effort? Is there any point in continuously refining our ideas or pursuing our ideals? Do we ever take time to ask what Great Wisdom has to say about life? What about the teachings of Shakyamuni Buddha, which have remained rele-

vant for 2,500 years since his times in India? Throughout
Buddha's eighty-year life span, he sought to convey to others
the significance and the purpose of life. He preached about the
Other World—which is, of course, the real world. Was everything he said untrue, calculated simply to mislead people? Of
course not. His was the teaching of a man of great enlightenment.

You may well be proud of your modern sophistication—
but do you know anyone today who has achieved sufficient
knowledge of the Truth to be able to gainsay any of Buddha's
teachings? What about the Truth as taught by Jesus 2,000
years ago? How can anyone claim that it is meaningless? With
more than one billion people still finding true wisdom in
Jesus' teaching, is it really possible that the God in whom he
believed was no more than a futile conviction or his own
deliberate invention?

Today it is a fact that many people regard "science" as
more important than teachings that have earned the veneration
of men and women for thousands of years. Unhappily, this
view is actively encouraged by a small number of scientists
who wish to be identified with the prestige and authority
attributed to science. Some consolation can be found, perhaps,
in the thought that this minority cannot in any way compare in
stature or character with the great teachers of wisdom. To
those people who adopt an attitude of pseudo-scientific ration-
alism, saying that they would believe in spirits if they could
see them, all that can be said is that it is a pity their outlook is

so limited. Limitations like this can be overcome, and worthwhile knowledge can be attained, just by meditating on the work of those great spirits who have undergone incarnation and become humans, such as Jesus or Buddha. Surely this is the very basis of true "scientific enquiry."

In practical life we touch the Buddha consciousness at times when we experience extreme emotions—when we are in love, or when there is a birth, a death, sickness, or disillusionment. The Buddha consciousness is inherent in the things of this world and in the things far beyond it. In this book, one way or another, I shall be exploring other aspects of Buddha and relating my exploration to the purpose and mission of life itself.

3. Existence and Time

One universal law holds good over all the numerous forms of existence on this planet—all the modes of organic life, all the inorganic objects of nature. This is the law of constant change. Everything in this world—humans, animals, plants, microbes, even minerals—is governed by this law of constant change; everything is in a state of flux. To rephrase it very simply, everything on the planet comes into existence, goes through a period of growth, slowly declines, and finally disappears. The human being, for example, is born, grows to adulthood, ages, and eventually dies. The law applies universally, both to natural objects and to objects made by human hands. It is thus true even for a motorcar: the car is constructed, for a time gives reliable service, begins to break down, and ultimately is scrapped. For plants the process is identical: a seed is scattered and germinates, a shoot appears, grows, and produces a flower; the flower wilts and vanishes, leaving just seeds or a similar initial form of life (like a bulb or corm) to mark its passing.

Forms of organic life (and inorganic existence) in our three-dimensional world have to pass through an identical system of four stages: birth → growth → decline → death. Static existence is impossible. Indeed, it could be argued that all things exist only on condition that they do remain in a state of flux. They are subject to, and contain within themselves, the characteristic of transmutation brought about over time. In a way, this might be said to resemble a piece of film shown through the projector of time. If we look at the cells that make up our body tissues, for example, we can see that none of us is the same person we were yesterday—but although our body cells change by the day, there is nonetheless some essential quality that unifies the cells, that makes the person the same unique person he or she has always been.

So, there must be something immutable and unchanging, in the inherently mutable and changing experience of existence in the flow of time. Whatever it is, it occurs in humans, animals, and plants. For instance, there is something that makes a flower a flower, rather than a random collection of vegetable cells. Even as a random collection of cells, it might be a flower one day, but the law of constant change would dictate that it mutate fairly definitively into something else the next. Yet a flower remains a flower. Yesterday, today, and tomorrow it can change only in the way a flower changes: it cannot transform itself into, for instance, an animal or a rock. A chrysanthemum cannot even turn into a tulip, and a tulip cannot become a daisy. A tulip is a tulip for all of its life.

Some element within the paradigm of constant change does not change. Something in the totality of what transmutes does not transmute. This something might be described as "real existence," the ultimate truth. There is an expression in Buddhist philosophy that sums up what I am getting at, albeit as a paradox: *Matter is void, void is matter.* It reveals real existence as something unchanging within a scenario of change, and in fact goes further: it proclaims that everything in this world of change is but a projection of an omnipresent Existence.

There is more to us than a mere compilation of changing body cells collectively known as a person. The true nature of a human being may be described not as subject to constant change within the flow of time, but as an essential, eternal, immutable real existence that corresponds also to "soul" and "spirit." By "spirit" I do not mean some mysterious phenomenon but the true essence of humanity: whatever it is that does not change in existence—individual life in its conceptual form. It is this spirit or essence that provides the context for the individual intelligence that inhabits and controls the body, the individual consciousness that informs the personality.

4. Finite and Infinite

We have briefly touched on the human experience of time and earthly existence, but there is a concept that goes beyond both time and space, namely the infinite and its relation to the finite. Is human life finite or infinite? Is the universe finite or infinite? All of us ask ourselves these questions at one time or another in our lives—but before I give an answer I would like to tell you a story.

Once upon a time a giant turtle lived on a beach. He was so big that it took a full ten minutes for him to move his right foreleg forward, ten minutes to move his left foreleg forward, and ten minutes more to move each hind leg. One complete movement involving all four legs thus took no less than forty minutes altogether.

There came a day when the turtle wondered if there was any end to his beach, and made up his mind to explore the extent of his world. Gazing straight down the length of the shoreline, and unable to see any limit to it, he summoned up all his strength and set his limbs in motion. Even though each complete forward movement of his body took forty minutes, he decided to start by surveying the waterline. He knew that if he left his footprints behind him in the sand he would be able to tell if he passed the same place a second time—from which it would seem that he was a very wise old turtle. On and on he plodded day after day, but the beach just extended forever in front of him. Finally, many, many days later, when he had used up all his reserves of strength in his exhausting pilgrimage, he died. But he died happy in the thought that by then he must surely have explored at least half of the world.

The next day, a fisherman who made his living on the vast sea circled by the beach and came across the turtle lying by the sea, and dragged it back to his house on the other side of the island where he ate it. You might think that it would have taken a long time to get the turtle all the way back across the island, but the young man was fit and it only took him ten minutes. There was a factor that the poor old turtle had not taken into account: the Pacific Ocean had washed away his footprints behind him, and he had been walking round and round an island that was in actuality really tiny.

When I consider the finite and the infinite, I cannot help but reflect on the story of the turtle and the fisherman. What was the most significant difference between the two? There were many differences, of course—in their walking speeds, in their overall size and shape, and in the range of their experience. But, to me, the basic difference between them was one of perception. The intention, enthusiasm, and perseverance of the turtle were all admirable; so why did he come to (what from our point of view is) such a sad end? It was because of the fundamental difference in their abilities to comprehend, to grasp matters in a wide enough scope of understanding.

Now, suppose instead of the turtle and the fisherman the same story was told about a contemporary materialist and someone who has mastered the Truth. I am sure some of you will feel offended and say, "I am not a turtle!" There are people today who firmly believe that life lasts only for a mere sixty or seventy years, and that everything ends at death. There are others who believe only in what they can perceive for themselves, and who refuse to acknowledge anything beyond their own five senses. Both these groups of people are no better than the turtle who set out to discover how large the world was; they rely on their own footsteps to show them where they have been, but end up going round and round in circles. Just as tragically, they believe they are living their lives to the full, whereas in reality they are doing no more than traveling around a closed loop.

We humans have been given eternal life. We are born time after time in this world in order to accumulate wisdom, the wisdom that comes with spiritual exercise. To grasp this is to approach closer to an understanding that the human life force is not limited in its activity just to the three-dimensional world. We belong, ultimately, to what I call the "real world," which extends way beyond the three dimensions we are all familiar with, to a fourth, and other dimensions beyond. These dimensions should be thought of not as having physical properties and boundaries but rather as being worlds in themselves, each having specific characteristics in terms of consciousness, as I will explain.

We can each live on any of a number of different planes depending on the level of our spiritual awareness. Such planes, or levels, are fundamental, for even the answer to the question of whether the universe is finite or infinite depends on which level we are talking about.

To use an analogy, and liken the universe to the human body and the various layers of clothing that can be put upon it, the everyday world of the third dimension may be held to correspond to the completely naked figure. The fourth dimension then corresponds to the undergarments, the fifth to the shirt that overlies the undergarments, the sixth to the sweater on top of the shirt, the seventh to the suit that covers the sweater, and the eighth to the full-flowing coat that envelops the suit. The ninth dimension might represent the hat upon the figure's head.

The analogy may be somewhat simplistic, but it represents the structural relationship of the several dimensions in an appositely accurate way. The lower dimensions are completely enclosed by the higher ones while remaining in close contact with each other. The higher dimensions resemble the lower ones but have a higher purpose.

5. The Multi-Dimensional Universe

Layers of clothing is one analogy by which to try to describe the multi-dimensional universe, but I want now to describe it without using an analogy at all. The description may seem both highly theoretical and dauntingly complex, but bear with me if you can: unhappily, it is always very difficult to describe aspects of the spiritual in terms of the physical and material.

We must first define more closely what we mean by "dimensions." The everyday world in which we live is often said to be three-dimensional. But what does that mean? A dimension is a mode of existence; the number of dimensions denotes the directional potentialities inherent in the mode of existence. For instance, a world that had one dimension only (the first dimension) would consist simply of points, each point on the same straight axis. If there were inhabitants of the first dimension, the only distinction between them would be the length of the straight lines they make up together as points, end to end.

The second dimension is the world of length *and breadth*, both of these properties being visible on a single (flat) plane. If there were any inhabitants of the second dimension, they would resemble shapes and figures on a flat piece of paper— they could be any shape at all, but they would have no depth. It also means that two "inhabitants" that had the same mathematical definition would be indistinguishable from each other.

The three dimensions of the world in which we live are length, breadth, *and height* (or depth). The addition of another dimension not only makes for concreteness in perceivable form but adds an immeasurable potential for variety in that form; two people are not identical in form unless they have exactly the same width, the same height and the same volume—a combination that is mathematically far more complex than any combination in the second dimension. It is also the case in the third dimension that things that exist in the same space also exist in the same time—but this is not so in the fourth dimension.

In the fourth dimension, the element of *time* is added to those of length, breadth, and height. For two objects to come into contact with each other (or for two people to shake hands) in the third dimension, they both have to exist at the same time on the same day of the same week in the same year. But this is not true in the fourth dimension, where two people shaking hands might not belong to the same era at all. For instance, in the fourth dimension it would be perfectly possible for a person from the present day to shake hands with someone from the twelfth century. Similarly, it would be impossible to tell whether the building in front of you genuinely existed in the present day or whether it was the "continued actuality" of a building that had stood there in the past. Even if it was only a simulacrum from the past, it would look and feel real; it would be there.

In the fourth dimension, everyone's watch would show a different time, and you could meet a woman who was born a thousand years before but who still looked young enough to be in her early twenties. It would not just be what had already happened that would be around you, for in the fourth dimension it is possible for events that are to occur in the third dimension's future to appear as vividly and as concretely as if they were happening now.

In the fifth dimension, the element of *spirituality* is added to those of length, breadth, height, and time. Inhabitants of the fifth dimension thus are distinct from each other not only in their shape (that is, their height, width, and volume) and their experience of time, but also specifically in their degree of spirituality. They would not be there at all if they had not reached a fair stage of advancement in spiritual understanding, certainly well past the realization that there is more to human beings than simply the material body. The main measure of spirituality is "goodness," and the fifth dimension may, therefore, be described as the realm in which the good assemble.

The sixth dimension adds a *knowledge of Truth* to the other factors of height, width, depth, time, and spirituality. The inhabitants of this domain thus differ from each other in all the aforementioned ways, and additionally in the extent of their knowledge of Truth. Again, to be in the sixth dimension at all, the inhabitants have to be morally good and to have an intimate knowledge of what we earlier called Buddha's Truth. The actual degree of the knowledge of Truth naturally varies from person to person, which gives rise to finer levels within the sixth dimension. But there is no one in the realm who is not familiar with, and an adherent of, the Truth.

6. The Higher Dimensions

In the seventh dimension and beyond—in addition to the

aspects of height, width, depth, time, spirituality, and the knowledge of Truth characteristic of the lower dimensions—is added the quality of *altruism:* freely being of service to others. In a sense, it could be said that the inhabitants of the lower dimensions all live egotistical lives, although I do not mean this in a necessarily negative way. Even in the highly moral sixth dimension, the inhabitants want something for themselves, if only more knowledge of Truth to improve themselves.

To resort to analogies again, some idea of the order of the higher dimensions might be suggested if we were to say that up to and including the sixth dimension, all the inhabitants correspond to students who have yet to graduate. Indeed, to pursue the analogy a little further, the inhabitants of the sixth dimension might be described as representing university students, those of the fifth dimension senior high school students (sixth-formers), of the fourth dimension lower school students, while we in the third dimension are still in elementary school.

It is only once they reach the seventh dimension that people there can finally consider their education complete and set out to become a true member of spiritual society. Their major preoccupation then is altruism, to be of service. Their hearts are filled with love, and everything they do is for the benefit of others. Loving and helping each other in their own dimensions, they also devote themselves to guiding the people of the lower dimensions, particularly assisting those in the fourth dimension who are disoriented after having left their physical

bodies behind. On occasion, members of the seventh dimension even allow themselves to be born again in physical bodies in the third dimension in order to live lives of love and
service to others. As this proves, the inhabitants of the seventh
dimension have already reached an exalted plane.

The factor that demarcates the inhabitants of the eighth
dimension—on top of the properties of height, width, depth,
time, spirituality, knowledge of Truth, and altruism—is what
in English is generally called *compassion.* The meaning of
"compassion" here is more than the dictionary definition of
"pity" or "sympathy." Here it is a very real empathy and
includes the desire to give. Such exalted beings freely and constantly give without discrimination—and this is the true form
of compassion. While the altruistic love of the seventh dimension is an act of giving love, the love of the eighth dimension
is even higher, and should be thought of as infinite love, an act
of continuously giving.

The love of the seventh dimension is still a result of human
endeavor; it is the bestowing on others of the love that a person has managed to accumulate in his or her work. The love of
the eighth dimension, however, is like the power of the sun—
it is inexhaustible. This is true compassion. The love of the
seventh dimension needs a subject, and the quantity of love
bestowed varies according to the subject. The love of the
inhabitants of the eighth dimension, on the other hand, is totally impartial; there is no room in this love for the petty discrim-
inations of the human heart. Because of their unstinting supply

of love, the inhabitants of the eighth dimension are truly qualified to lead others.

In the ninth dimension the element of the *universe* is added to the properties of height, width, depth, time, spirituality, knowledge of Truth, altruism, and compassion. The people who live in the dimensions up to and including the eighth dimension exist in realms that envelop and are influenced directly by the energy of the Earth. The ninth dimension, however, is not restricted solely to the Earth but is linked with the spiritual worlds of other planets beyond the Solar System. The dimension's inhabitants may, therefore, be described as offering guidance to the inhabitants of the terrestrial spirit group from the midst of the evolving universe. Most of those who are revered as the "fundamental God" in any major world religions are inhabitants of this ninth dimension. They are the source of all law, and may be said to differ from each other only in the color of the light through which they express the law. There is but one Universal Law or Law of Buddha, but it can be expressed through any of seven colors that each correspond to the character of the inhabitant of the ninth dimension who is interpreting it.

Beyond the ninth dimension lies the highest level which the terrestrial spirit group may attain, the tenth dimension. No inhabitants of this realm have ever taken human form on the Earth, for it is inhabited only by the three consciousnesses. The special properties that distinguish the tenth dimension (above the empathetic compassion of the eighth dimension and the cosmic universality of the ninth dimension) must be regarded as *creation* and *evolution*. The inhabitants of the tenth dimension no longer have individual personal attributes like human beings; they differ only in the roles that they have in relation to creation and evolution.

The three consciousnesses comprise the "Grand Sun Consciousness," the "Moon Consciousness," and the "Earth Consciousness." The Grand Sun Consciousness governs the positive and active force within all life on Earth, including human beings; the Moon Consciousness governs the passive, refined, and feminine aspects; and the Earth Consciousness represents the consciousnesses of the Earth as a life force, and has been responsible for the formation of all things on the planet. Everything that has happened during the 4,600-million-year history of the Earth has resulted from the operations of these three consciousnesses.

Although the terrestrial system extends only as far as the tenth dimension, the Solar System also has an eleventh dimension. The characteristic property of this dimension is the *mission of the sun*, comprising the Sun itself as both a life force and a spiritual being.

The twelfth dimension is the Galactic Consciousness, an extremely powerful spirit that is responsible for planning our galaxy and that controls hundreds of thousands of stellar consciousnesses of the eleventh dimension (as opposed to the planetary consciousness of the tenth dimension).

I can only hope that the words I have used in this brief

summary of the dimensional layout of the universe have been clear enough for readers to grasp. Twelve dimensions and their inherent properties and entities may seem inordinately complex, but I am obliged to go on to add that the Primordial Buddha, the highest consciousness—the Primordial God of the universe is thought to inhabit a realm in the twentieth dimension, or even beyond that.

7. The Birth of the Stars

One of the major roles of religion is to try and explain what happens to a person after death, whereas a major role of the natural sciences is to try and fathom the mystery of the origin of life, and explain how life in its various forms first came into existence. Neither of these viewpoints fits precisely with what I now want to say about the birth of the stars and the origins of life. I mean instead to show that, in some respects, the aims of both religion and science are one and the same. You may well be astounded by what I have to say.

Three-dimensional space—of which the Earth we live on is a part—is said to have come into existence some 40,000 million years ago. In the language of dimensions, the Buddha of the Grand Cosmos, the Primordial God, exists in the twentieth dimension or higher, and has existed as a conscious being for several hundred thousand million years at least, from a time that is unimaginably distant.

This Supreme Consciousness conceived the creation of the three-dimensional universe approximately 100,000 million years ago. Around 20,000 million years later, or 80,000 million years ago, He used His will to create the great spirit who would control the three-dimensional universe. This was the origin of the consciousness of the thirteenth plane, and the birth of the first spirit in the recognizable universe. The Cosmic Spirit of the thirteenth dimension is a projection of the supreme consciousness and its task was to create the Grand Cosmos.

Approximately 65,000 million years ago, the Cosmic Spirit of the thirteenth dimension created the nebular consciousness of the twelfth dimension. No fewer than two million of these nebular consciousnesses are believed to exist, and the Galactic Consciousness—of which we are a part—is one of them.

Around 60,000 million years ago, the Spirit of the nebular consciousnesses of the twelfth dimension created the stellar consciousnesses of the eleventh dimension—this was the origin of eleven-dimensional space. The Spirit of the Galactic Consciousness of the twelfth dimension created the Spirit of the Solar System in the eleventh dimension.

Next, about 53,000 million years ago, the stellar consciousnesses of the eleventh dimension in every galactic system began the creation of planetary consciousnesses, heralding the birth of the tenth dimension of our Solar System. The planetary consciousnesses of Mercury, Venus, Earth, Mars, Jupiter, and Saturn were born. The creation of these planetary consciousnesses was completed 42,000 million years ago.

Approximately 40,000 million years ago, an extraordinary phenomenon occurred within the Spirit of the Grand Cosmos in the thirteenth dimension. Something resembling nuclear fusion and fission took place within its body, creating a cosmic fireworks display of cataclysmic proportions—which we now refer to as the Big Bang. In an instant of time, three-dimensional space abruptly appeared within the body of the Cosmic Spirit of the thirteenth dimension, forming something that we might compare with an organ in the human body, in that respect not entirely unlike a giant stomach. But, in composition, it bore little resemblance to the relatively systematic spatial domain we know today.

Therefore, to bring intelligible order to this cosmic chaos, the nebular consciousnesses of the twelfth dimension, the stellar consciousnesses of the eleventh dimension, and the planetary consciousnesses of the tenth dimension all worked together to gradually create the concrete forms of the nebulas, stars, and planets in three-dimensional space.

Over the 40,000 million years that have elapsed since the appearance of three-dimensional space, the rate at which nebulas and stars have been created has varied from time to time and from cosmic locality to cosmic locality. Our own solar system was born approximately 10,000 million years ago. The planet Mercury was formed 7,000 million years ago, Venus 6,000 million years ago, and Earth about 4,600 million years ago. But this is to use the terms of historical cosmology to account for the world as we know it, and it should be remembered that the stars and planets existed originally as independent conscious life forms. Readers may be unhappy at seeing such familiar terms of historical cosmology reinterpreted. Yet my view is surely worthy of some attention as it is the only one that combines what is taken to be scientific fact with a vision of the place and purpose of Supreme Consciousness.

8. The Birth of Human Spirits and Other Life Forms

Exactly when human spirits first came into being is not known. As we have seen, however, three-dimensional space first appeared in its primal form 40,000 million years ago, an event that was followed soon after by the birth of the nebular, galactic, and stellar bodies. So we can say that the first life was that of the stars, and the stars then served as a basis for the formation of all other kinds of life.

To avoid unnecessary complication, let us examine the origins of life just within our solar system. Our sun was born as a star in three-dimensional space approximately 10,000 million years ago. The first of the planets to form, 7,000 million years ago, was Mercury, but conditions on the planet made it unsuitable for life.

The first life in our solar system evolved after the formation of the mysterious and beautiful planet Venus some 6,000 million year ago. It was not for another 500 million years, however, that the ninth dimension manifested itself within the Solar System, 5,500 million years ago. A great spirit was created—a spirit with a highly developed character more active than spirits of the tenth dimension and therefore better suited to rule the life forms that were to appear on the planet. This great spirit of the ninth dimension, who embodied the character of Venus and ruled the planet, was El Miore.

The first three-dimensional life form that El Miore created on the planetary surface of Venus was experimental, representing a cross between animal and plant. Its upper half resembled a lily; its lower half comprised two legs, like those of a human being, on the back of which was a thick growth of leafy foliage to allow metabolism through photosynthesis. A highly self-sustaining life form, it additionally had the characteristic of being long-lived.

Next, El Miore separated plant and animal life, and left them to evolve for 2,000 million years. Both the plants and the animals that resulted from this were different from those on Earth in that they were graceful and beautiful. The plants put forth flowers like jewels that emitted ethereal perfumes; the animals were no less refined, and some were even capable of speech.

Eventually, El Miore created a race of Venusians who closely resembled humans on Earth today, and for more than 1,000 million years he sought to polish and refine them over successive generations. During this period, hundreds and thousands of civilizations rose and fell, until the race as a whole eventually became so sophisticated that it gained the ability to travel to the planets of other star systems.

In their final form the Venusians were very similar in outward appearance to modern human beings, but their intelligence was the IQ equivalent of more than 300, and both men and women possessed a celestial radiance and shone with an opalescent beauty. The women were particularly exquisite; in comparison with them the loveliest women on Earth would seem imperfect, if not downright simian. A paradise was created on Venus that overflowed with idyllic love, beauty, and intellect. Its inhabitants formed a Utopian society in which the watchwords were love, wisdom, self-reflection, and progress.

It was not to last. Just as the planet's inhabitants were approaching the stage of spiritual mastery, a very different prospect for the future became evident to El Miore through his link with the Supreme Consciousness.

El Miore was effectively informed that although the experiment on Venus had been a brilliant success and a state of perfect harmony had been achieved there, further spiritual developments would henceforward be very difficult. A massive volcanic eruption was soon to occur on Venus, making it problematic that any advanced life form would survive. Some of the inhabitants would, therefore, have to emigrate to other planets with which there were good relations, and be prepared to assist the local evolutionary process in their adopted environments. Those who did not emigrate would remain in the spirit realm pertaining to Venus for a few hundred million years in order to help in the establishment of a terrestrial spirit group on the neighboring planet Earth.

El Miore was instructed to start again from the beginning on Earth, and over time to create another utopia. New spirits should be called in from other star groups and educated to promote the development of the galaxy.

From this time on, it was the advancement of the Earth that became the main concern of El Miore. The Earth had formed 4,600 million years before, at a time when the experiment with life and the development of civilization was already well under way on Venus. The spirits of the tenth dimension duly pondered over the form life on Earth should take, and looked to the advanced experiments on Venus for guidance. Moreover, because conditions on Earth were better adapted for life, they concentrated on producing a more dynamic evolution when they came to form the terrestrial spirit group. Acting on the advice of El Miore, the three grand spirits of the tenth dimension—the Grand Sun Consciousness, the Moon Consciousness, and the Earth Consciousness—decided to lay down two basic laws for all life on Earth. First, different levels of development were to be assigned to the life forms created—high and low, or superior and inferior. Second, the life span of all creatures would be limited but subject to the principle of reincarnation, allowing the spirits to travel through the multi-dimensional worlds between lives.

In accordance with the first law, simple organisms like amebas and plankton were created on Earth as the basis of animal life, approximately 3,000 million years ago. Mold and other fungal growths appeared some 2,600 million years ago as the forerunners of plant life. Thereafter, more and more complex life forms were increasingly seen on the Earth.

The initial step in applying the second law was the creation of a low-level spirit world. This was later developed into the fourth-dimensional Posthumous Realm (as we know it now). But, at that time, it was not totally separate and resembled a misty veil covering the Earth. The basic creatures and plants of the time moved between it and the Earth in a very simple form of reincarnation.

About 600 million years ago, the three grand spirits of Earth realized that the time had at last come for them to create superior life forms on the planet. First, they fashioned the ninth-dimensional spirit world and invited El Miore from the more advanced planet of Venus to come and reside there as the first grand spirit of Earth to have human characteristics. El Miore began his task by transferring to Earth the early spiritual life forms he had created on Venus, then going on to create more complex life forms based on mammalian principles.

What exactly was the process he used to create these more complex life forms? First, he created the conscious form of an animal such as a mouse, a rabbit, a dog, or a cat in the low-level spiritual world. Then, he progressively materialized them on the Earth. By this means, complex life forms soon began to prosper on the Earth, and the system of reincarnation became firmly established. Perceiving this, El Miore approached the grand spirits of the tenth dimension and insisted that the time was now right for the introduction of humankind. The grand spirits concurred, and approximately 400 million years ago the human race was born.

At this time, El Miore—former ruler of Venus and the first being of the ninth dimension in the terrestrial spirit group—changed his name to El Cantare, a name that means "the beautiful land of light, Earth." It was the same El Cantare who entered the third dimension and took on human form some 2,500 years ago, appearing in India, where he came to be known as Gautama Siddhartha, the founder of Buddhism.

9. The Inception of the Terrestrial Spirit Group

In creating the terrestrial spirit group, El Cantare decided it should conform to the two basic laws of life. Firstly, human beings were to enjoy differing levels of consciousness and would be provided with a place in which they might evolve eternally. Secondly, while human life would be comparatively short, the souls would transmigrate between the Earth and the spirit world in an endless cycle of reincarnation.

Taking the highly developed spiritual life force of the Venusians at his basis, El Cantare started to create human beings. He generated a huge sphere of light made up of ninth-dimensional compassion and wisdom, and into it sent the most highly developed spirits of the Venusian race, granting them the power of regeneration. In due course, they divided up into a multitude of smaller spheres of light, which became the Guiding Spirits of Light of the eighth dimension of the terrestrial spirit group and below. There were hundreds of them.

In order to endow the Guiding Spirits of Light with individual characteristics, El Cantare used the full power of the ninth dimension to materialize them on Earth. At first, they appeared in the form of shimmering, translucent shadows. But as time went on, they gradually took on more human shape, until finally their bodies seemed to glow with radiance. El Cantare was delighted by the beauty and perfection of his creation. First five, then ten, then a hundred, then five hundred of them materialized out of the air. El Cantare divided them into two groups; based on the quality of the Venusians, he bestowed on those to his right the light of wisdom and courage, and on those to his left the light of grace and beauty—and this is the origin of the two sexes of humankind.

All were well developed in spiritual terms; they were among those who became the gods of ancient Greece and the masters of Buddhism. Their corporeal descendants steadily multiplied—many of the peerless Venusian spirits themselves also experienced life on Earth in human form—until their numbers eventually reached a massive 770 million.

Concluding that these descendants of the superior humans he had created ought to experience the responsibilities of leadership, El Cantare decided to introduce less-evolved beings—creatures a little more advanced than anthropoid apes—for them to teach and guide. He, therefore, resolved to import human-like beings from other planets, and contacted three ninth-dimensional grand spirits from distant constellations to ask them for their opinions on whom to bring in. The spirits he consulted were Amor (Jesus Christ) from Sagittarius,

Therabim (Confucius) from Cygnus, and Moria (Moses) from Cancer.

It was at this time that gigantic life forms such as the dinosaurs were beginning to prowl the land surface of the Earth, and there was some apprehension that colonists unused to the terrestrial environment might be in danger for their lives. For this reason, the race chosen to comprise the first immigrants was a warlike people from the Magellanic Cloud who were both extrovertly audacious and aggressively independent. Their technological expertise was relatively advanced, at least enough for them to be able to make the journey to Earth in their own spacecraft. In outward appearance they were very close to the people of today, except that they had pointed ears and a tail much like that of a cat. These features were gradually to disappear through the process of evolution, although some of the people, on returning to the spirit world, reverted to their original form as a result of residual racial memory.

The superior beings originally brought into existence on Earth by El Cantare—who may be thought of as something like Earth's first royal family—worked to find ways to help the newcomers assimilate and acclimatize to terrestrial conditions. Among the immigrants' leaders, however, was a group who, although they were possessed of a high degree of light energy, behaved like arrogant demigods and upset the equilibrium of the population. El Cantare dealt with them by banishing them to a minor realm he had created on the reverse side of the major realm of heaven—and this is how the minor realms in the sixth, seventh, and eighth dimensions came to be established.

The leader of the inhabitants of the minor realm in the ninth dimension was Enlil. His lieutenant, Lucifer, took human form on Earth some 120 million years ago under the name Satan, but became so enamored of the status, renown, material possessions, and sensual pleasures available to him on Earth that he fell from grace. Unable to return to the higher spiritual world, he founded the Realm of Hell on one of the lower planes, and inaugurated a rebellion. Since then, he has acted as ruler of hell and became known as Lucifer.

Perceiving that the immigrants from the Magellanic Cloud were overly self-centered and contentious, El Cantare decided to introduce another group of humans to the Earth. Then, 270 million years ago, 1,000 million people arrived in a huge fleet of space ships from the constellation Orion: the second immigration from space. By this time, the number of spirits in El Cantare's Venusian spirit group that had experienced life on Earth had reached 10,000 million, so they were able to absorb even this large a number of immigrants.

Three great spirits of the ninth dimension named Achemene, Orgon and Kaitron also took advantage of the migration to visit the Earth. Achemene is also known as Manu, the "father of man" in Indian mythology. Orgon, also called the Master Maitrayer, was very active during the age of Lamudia or Atlantis, but has not incarnated on Earth much over the last 10,000 years. Kaitron, known to Theosophists as Koot Hoomi, is associated with science and technology and has taken on human form twice on Earth: once in ancient Greece, where he was known as Archimedes, and more recently in Britain, where he called himself Isaac Newton.

In the face of a huge influx of souls, the fifth dimension—the Realm of the Good—was prepared and expanded to accommodate them. Approximately 150 million years ago, El Cantare himself became incarnate on Earth and founded a magnificent civilization of light. He established Buddha's Truth (and all that it entails) on the Earth, which led to great advances in teaching people from other constellations. Vast numbers of people were converted to a belief in El Cantare, and a sense of unity grew among all the inhabitants of the planet.

El Cantare's spirit group continued to expand through repeated spectral diffraction in the upper spiritual world. By 130 million years ago, the spirits numbered in excess of 40,000 million—and to celebrate this fact, it was decided to invite a further 2,000 million people to migrate to the Earth from the Pegasus star system. During this migration, the ninth and tenth great spirits of the ninth dimension, Theoria and Samatria, came to Earth. Theoria took human form on Earth some 3,000 years ago. Known as Zeus, Theoria was later worshipped as the god of that name in Greek mythology. Samatria became incarnate on Earth twice in the region that is now known as Iran. Given the name first of Zoroaster and then of

Mani, he successively became the founders of Zoroastrianism and Manicheism.

In this way, ten grand sprits assembled in the ninth dimension, where they established a system for the guidance of the terrestrial spirit group. It was around this time that the spirit world of the fourth dimension was set aside to cater for the new migrants to Earth.

10. The Expansion of the Terrestrial Spirit Group, and the Appearance of the Corrupt

As we have seen, by 130 million years ago El Cantare's spirit group had expanded to more than 40,000 million, and the immigrant population from elsewhere in the universe had reached some 3,000 million. Nonetheless, Enlil—one of the great spirits of the ninth dimension—proposed that there should be a further large increase in the number of spirits from other star systems. He suggested in addition that more efficient spirit training could be achieved if each spirit of the relatively advanced immigrants could be split up to create five people, who could then be sent down one after another to live on Earth.

The light energy of the higher dimensions found a way to achieve this, but many of the spirits produced in this fashion were inferior and, while living on Earth, tended to forget their original form as spirits, so abandoning themselves to materialism and worldly passions. They exerted an increasingly bad influence on the more morally sound spirits, and after they died they began to create a field of their own in the lower dimensions. Groups of people with dark thoughts gathered in the spirit world of the fourth dimension. This is the origin of hell.

This was the second fiasco for which Enlil was responsible (the first being the disruption he caused at the time of the original migration). El Cantare duly addressed him in terms of stern correction.

Much worse, Lucifer's rebellion against the high spirits of heaven 120 million years before, which had created the vast Realm of Hell, caused dark clouds formed by the conceptual energy of the spirits in hell to billow out, cutting off the light of Buddha (God) and making it a cold, bleak world. This had repercussions elsewhere, for the presence of this dark world in the fourth dimension inevitably meant that some areas in the third dimension were in turn cut off from the light of Buddha (God). For the last 120 million years, therefore, hell has thrown its shadow over the Earth, giving rise to all kinds of evil and malign chaos.

Another result of all this is that for the last 100 million years, the Masters of the upper realms have had to struggle to save the world of the third dimension from demons and malicious spirits, led by Lucifer, who are trying to expand into the third dimension to escape from the agonies of hell. To this end, El Cantare himself has repeatedly sent parts of his own consciousness down to Earth to create a powerful system of training to foster guiding spirits of light.

This book, *The Laws of the Sun,* has been written in an effort to enable the Sun of Truth once more to illuminate this realm of the third dimension. Succeeding chapters will reveal how important these laws are, and why it has been necessary first to include a description of the origins and great significance of the terrestrial spirit group. For the laws represent a means of restoring the original world of light, the Land of Buddha, and are a means, too, of winning salvation.

Two: The Truth Speaks

1. The Truth about the Soul

Chapter One described how the universe was created and the terrestrial spirit group came together. According to the evidence, what happened was that spirits in higher dimensions formed beings in lower dimensions. The grand spirits of the higher dimensional planes were themselves brought into existence by the will of the Supreme Consciousness—Primordial Buddha. Following the appearance of the stellar and planetary consciousnesses, a disturbance in the fabric of the Grand Cosmic Spirit occurred that led to the formation of the third dimension. Soon after this, the stars and star clusters were created, and in each solar system a life sphere for human spirits was established in the ninth dimension and below.

In our solar system, the spirit world enveloping the Earth began with the creation of the Cosmic Realm in the ninth dimension. This was followed by the appearance of the Realm of the Master (Nyorai, or *Tathagata* in Sanskrit) in the eighth dimension, sometimes alternatively called the Diamond Realm. Then came the Realm of the Angels (Bosatsu, or *Bodhisattva* in Sanskrit) in the seventh dimension, the Light Realm in the sixth dimension, the Realm of the Good (or Good-hearted) in the fifth dimension, and the Posthumous

Realm in the fourth dimension (encompassing both the Astral Realm and the Realm of Hell).

Structures involving dimensions that are similar and correspond to the dimensions pertaining to Earth naturally exist in other parts of the universe. But, whereas the realm of the ninth dimension is linked with those throughout the universe, those of the eighth dimension and below differ in that they belong to the spirit world of the individual planets and develop independently of each other.

What we refer to as a "soul" or "spirit" is an expression of the Supreme Consciousness (Primordial Buddha) of the highest realm, manifested in the lower dimensions. This means that Primordial Buddha is not a separate entity who exists apart from us, but is the high-dimensional consciousness from which we obtain our existence. In reality, then, we are all part of this consciousness, part of Primordial Buddha's expression of self.

Primordial Buddha created the Grand Cosmos and everything that has life within it as individual facets of His own self. All Creation is the expression and reflection of the consciousness of Primordial Buddha. Were Primordial Buddha ever to decide that its continued existence was undesirable, the seemingly infinite universe of the third dimension would disappear in an instant. So ephemeral is our existence that if Buddha were to abandon His desire for self-expression, human beings would from that moment cease to exist as a life form. Insignificant as we may appear, however, our individual lives are nonetheless part of Buddha's consciousness—and in that respect we may consider ourselves a very advanced form of being.

As a part of Buddha's own expression of self, indeed, we may each take pride in ourselves and act with total self-confidence. This is one of the true benefits of the soul; the religions and advanced philosophies of times gone by have been developed and passed down with the sole object of enabling people to understand this truth. It is the ultimate aim, too—although not yet obvious—of natural science and space technology, both of which have made tremendous strides forward in recent years, that this truth about the soul should be revealed.

It is a wonderful and inspiring thing to know that we are a part of the great Buddha Consciousness. With that knowledge firmly in our minds, I would now like to go on to describe the ideal state of the soul, the ideal existence of humans as living beings—and as I do so, surely more of the Truth will be revealed.

2. The Nature of the Soul

The human soul is a part of the Buddha consciousness. By means of the profound study of the soul it should, therefore, be possible to obtain a glimpse of the true nature and character of that great consciousness.

The soul has several distinctive features, the first of which is its power to create. The soul has been granted the ability to transform itself at will—in other words, as a self-aware individual you are free to decide what sort of thoughts you have. For instance, you can practice love or freedom to the highest degree. You are free to control the quantity of light within yourself. You are free to increase it and by so doing raise yourself to the highest dimension of existence, and alternatively you are free to reduce it and so drop to a lower level.

But is it part of the soul's nature to think and do evil, to corrupt itself? Is it part of the soul's capacity for creation to fall from grace or to produce hell? The answer to these questions is both "Yes" and "No."

It is "Yes" for the very reason that the soul has been granted the freedom to create. It cannot be free if it has obstacles set round it or limits placed upon it. In that situation it is restricted. But the answer is also "No" because to do evil or to produce hell is not part of the original aim of the soul. Evil is not an inherent part of the soul's constitution. Evil is a distortion caused by friction between two souls both striving to pursue their own freedom. There is no evil that a person can commit if totally alone. Evil is something that first reveals itself in the presence of another person, another form of life, or another object.

Throughout history, arguments have raged back and forth about the dualist concept of good and evil. The basic questions are: "How can evil exist in a world that is the creation of Buddha?" and "Is evil a hidden aspect of the character of Buddha Himself?" It is obvious, though, that evil cannot be part of Buddha's character because evil is by definition "to obstruct the realization of Buddha's plan." Evil is nothing more than a temporary distortion or warping of the world of the mind or the world of phenomena brought about by conflict between two people simultaneously pursuing the freedom given to them by Buddha. It has no basic existence of its own but comes about through the functional and behavioral interaction of others.

The second characteristic of the soul is its ability to focus the light of Buddha and act as a medium for its emission. I should explain what I mean by "the light of Buddha" in greater detail. It is the energy of Buddha that fills the Grand Cosmos and shines down on the fourth dimension and those dimensions above it, just as the sun shines on our world. Just as we cannot live without the energy of the sun, so the beings of the Real World in the fourth dimension and above cannot exist without the energy they receive from Buddha.

The soul has the ability to focus, absorb, radiate, and amplify the light of Buddha. People who have the power to absorb and radiate large quantities of Buddha's light are full of light. They are the Guiding Spirits of Light, the Nyorai (Tathagata, or Masters) and the Bosatsu (Bodhisattva, or Angels) who have the ability to focus and transmit vast quantities of the light of Buddha. Throughout their lives, people's souls absorb and transmit the light of Buddha, but the Guiding Spirits of Light have the capacity to supply other people with the light of Buddha, to give light to the world and to brighten people's hearts and minds.

Every soul has the power to absorb and transmit light, but spirits in hell have been cut off from the power of the light of Buddha. They have created for themselves a cloud of evil conceptual energy that lets no light through—it is as if they live in a dank, dark cave. They no longer rely on Buddha's energy to support their lives, preferring instead to live off the evil conceptual energy that lurks in the minds of people on Earth. They come to steal this store of energy, to tap into the dark recesses of people's minds, to suck off the energy and lead the people astray. They are like vampires, siphoning off the life force of living men and women.

To avoid being possessed by these spirits, we have to prevent them from tapping into our energy. In other words, we must avoid creating dark, dank regions in our minds; we must prevent the formation of any such mental recess that may grow like a cancer and cut us off from the light of Buddha. If we can do this successfully, the inhabitants of hell will be deprived of their only source of energy, and hell itself will cease to exist.

3. The Incarnations of Buddha

I must now explain what I mean by the term "the Bosatsu of Light." The word *Angel* has distinctly Christian overtones, just as the word *Bosatsu, or Bodhisattva*, has Buddhist implications. But, in fact, some of the Nyorai, or Tathagata, in Buddhism correspond with Christian archangels, and the meaning of the word *Bosatsu* certainly contains connotations of the angelic. As stated previously, however, both Christianity and Buddhism are teachings of the Truth, the only difference being in the color of the light brought about through the individuality of their founders. It makes no difference if these spirits of the high realms are known as Great Guiding Spirits of Light or as Angels of Light. They represent the incarnations of Buddha.

But why do they exist at all? What was Buddha's purpose in creating these higher spirits that took on human form? After all, Buddha is said to have created all people equal, but the very existence of such high spirits would seem to indicate some kind of discrimination between outwardly similar facets of Buddha's self-expression. Can it imply that ordinary people are doomed to live ordinary lives while higher spirits live on a higher level?

In order to understand why there should be this division between higher- and lower-level spirits, it is necessary to understand that Buddha looks at the world from the viewpoint of "equality" and "fair judgment." Everything—human, animal, plant or mineral—contains within it a measure of divine nature. No matter how it is manifested, it is all an expression of Buddha's will.

To put it another way, all life and all things represent the diamonds of Buddha's wisdom. Buddha scattered these diamonds in their various types throughout the universe to produce humans and animals, and by doing so created the artistic beauty that is life. Buddhism perceives these diamonds as "Buddha-nature," and holds that they reside in all things and all people as the creatures of Primordial Buddha. It makes no difference, therefore, if a being is a higher spirit or a lower spirit—all are equal in that they embody the life force of Buddha. Any suggestion of "discrimination" stems from our own use of the words "higher" and "lower" in the terminology that describes their different states.

The only difference between the two types of being is the degree to which they have developed. There are highly developed spirits, developing spirits, and undeveloped spirits, but all are walking the same road; some have merely progressed farther than others. The Great Guiding Spirits of Light are very ancient souls, and have accordingly gone on ahead and have approached much closer to the goal of the state of Buddhahood. Most of the undeveloped spirits, on the other hand, are in the state they are because not long has passed since their souls were created. Still young, they inevitably trail behind those who are older. Is that unfair? Is it unfair to put differential values on the distances people have traveled in their journeys?

Instead of considering equality and inequality, we should rather consider achievement—what have people achieved? For just because spirits might be ancient does not necessarily mean that they have progressed far down the road. Some turn back and retrace their steps down the path they have come. There are those who, despite having once been angels, are now demons in hell. At one time they had progressed a long way down the road, but they then decided for one reason or another to turn back. These could be called "regressive spirits" rather than simply undeveloped spirits.

All spirits are equal in that they all travel the same road toward Buddhahood, but they should at the same time be judged fairly on the actual distance they have managed to cover. The higher spirits, the incarnations of Buddha, have been awarded tasks in proportion to their achievements in the past. All the lesser spirits are in their turn eternally striving to try to attain a state similar to that of the higher spirits.

4. The Anatomy of the Soul

The different levels of development (or achievement) on the part of spirits (or souls) thus result from Buddha's views on equality and fair judgment. But a word about the anatomy of a soul is now required. It is often said that there are "core" spirits and "branch" spirits—that the soul who comes down to Earth as a person represents only a superficial consciousness, whereas a deeper consciousness remains behind in the Real World. These concepts can be clarified.

In the beginning, Primordial Buddha (in the twentieth dimension or even higher) created the Grand Cosmic Spirit, the consciousness of the thirteenth dimension. This spirit in turn created the twelfth-dimensional consciousnesses of the nebulae, who then produced the stellar consciousnesses of the eleventh dimension. The stellar consciousnesses were thereafter responsible for creating the planetary consciousnesses of the tenth dimension.

The first consciousness with human characteristics appeared in the ninth dimension. Although these great spirits of the ninth dimension have individual character, they are made up of too much energy to be contained within a human body, and when they take it upon themselves to assume human form in the third dimension they are obliged to use only a small part of their consciousness to do so. When the great spirit of Gautama Buddha or Jesus Christ came down to Earth, for instance, only a part of its being was dispatched to create the individual soul that became the person. From this it is evident in this case that the soul is a spirit that has an individual human character and that, when it leaves the body to return to the ninth dimension, it becomes a part of the memories of the great spirit. It is possible for one of the great spirits of the ninth dimension to split itself into an infinite number of souls at will.

However, the situation is different in relation to the Great Guiding Spirits of Light from the Nyorai (Tathagata) Realm in the eighth dimension. They too are great spirits, but they have a more developed and individualistic character, and generally remain in heaven in a complete (i.e. undivided) form. All the same, if the situation demands, the spirits also can segregate themselves into any number of souls. An example of this is Yakushi Nyorai, a master of medicine, who for most of the time remains in its integral form in the eighth dimension, but who, when the circumstance calls for medical aid, can pour out its light to create thousands or even tens of thousands of entities who can each help guide the spirits in heaven or the people on Earth. In general, then, it is only in emergencies and for a specific and temporary purpose that the Nyorai of the eighth dimension leave their normal entire form and divide themselves into whatever number of spirits is required. In this respect, they differ from the great spirits of the ninth dimension who, although they each have only a single characteristic light, are capable of creating a sufficiency of human souls for multiple tasks.

The spirits of the Bosatsu (Bodhisattva) Realm in the seventh dimension have quite individual characters, and there is good reason for this. Unlike the spirits of the eighth dimension, some of whom have never visited the surface of the Earth, all of the spirits of the seventh dimension have accumulated experiences of life as humans on the planet.

The seventh-dimensional spirits of the El Cantare spirit group attach great importance to teamwork; as a rule they operate in basic groups of six at a time. The leader of the group is regarded as the core spirit; the other five members as branch spirits. Each of the six, in turn, comes down to Earth to carry out the work of the Bosatsu (Angels), while the spirit whose turn is next takes on a monitoring and supervisory role as "guardian spirit" for the spirit actually on the Earth. The guardian spirit is thus provided with an opportunity to study mortal life. Society has become so complex, however, that it is increasingly common for the spirit just returned from incarnation on Earth to take on the guardianship of the next spirit. The experiences of each individual spirit are shared by all, and all of them have a similar disposition. Indeed, the group of six spirits may be regarded as elements comprising a single spirit in much the same way that the human body may be said to comprise six parts with its head, torso, and four limbs.

In the Light Realm in the sixth dimension, the spirits rarely have a mind to act as a member of a group of six. Each individual operates independently, and for this reason has little understanding of such terms as "brother soul," "sister soul," "core spirit," and "branch spirit."

In the Light Realm and below live some spirits that were created some 100 million years or so ago, when light from the higher dimensions split each of the souls in the sixth dimension to form five more duplicate spirits. The spiritual level of these newly formed duplicates, however, was lower than that of the originals, and most of them in due course took up residence in the fifth-dimensional Realm of the Good or the fourth-dimensional Posthumous Realm. It is because these spirits have needed to regain their spiritual rank that humans have been transmigrating between this world and the spirit world for the last 100 million years.

In the sixth dimension and below, the core spirit is mainly involved in protecting and guiding the spirit undertaking spiritual practice on Earth. If ever the nature of events experienced on Earth is such as to cause serious deviation of the characters within the group of six spirits, threatening to disrupt the group altogether, the group is re-formed through the power of Buddha's light.

5. The Formation of Guardian Spirits and Guiding Spirits
Some people who are spiritually aware talk about "guardian spirits" and "guiding spirits," so an explanation of what those terms refer to may be useful. It is quite common to hear it said that everyone has a guardian angel (or guardian spirit), and that, if the guardian spirit is powerful, the person whose life is being monitored will have a happy and comfortable life. But if it is weak, the person will meet with misfortune.

Guardian spirits do exist, and there is one assigned to every person. It is also true—to some extent—that the power of the spirit has an influence on the destiny of the person. But how is it possible for guardian spirits that exist in the Real World to protect those living on Earth? This is what I want to make clear.

There were no guardian spirits when the highly advanced humans of the El Cantare spirit group first began to live on Earth more than 350 million years ago. In those days, the hearts of the people were so pure that it was possible for them to communicate directly with the spirits of the Real World. Moreover, at the time there was no hell and malicious spirits did not exist, so there was no need for guardian spirits anyway. But about 120 million years ago, disaffected spirits began to assemble on the lowest level of the Real World in the fourth dimension, gradually forming the dark Realm of Hell. Cutting themselves off from the energy of Buddha's light, they began to generate confusion among the inhabitants of the Earth, breeding thoughts of desire, negativity, and disharmony in the people to create the energy they needed to live on.

This was a development that was totally unexpected. When they saw what was happening, the Guiding Spirits of Light held an emergency meeting to discuss what they should do to resolve the situation. They eventually agreed on a plan put forward by Amor, known in this world as Jesus Christ, whose proposal was threefold. First, to prevent the malicious spirits from achieving total control over the Earth, humans would no longer be able to communicate directly with the Real World but would be encouraged to work toward achieving a better spiritual life in the material world. Second, all persons would be assigned a guardian spirit when born on Earth, to help protect them from the temptations of hell. Third, in order to stop the people from forgetting their origins in the Real World altogether, the Great Guiding Spirits of Light would be sent to Earth at regular intervals to preach religion and instruct them in the existence of the other world.

These three principles have been observed for more than 100 million years, but the Realm of Hell has grown so large that it has become difficult for a single guardian spirit to protect a person who has come down to Earth to undertake spiritual training—particularly in view of the fact that only religious leaders are allowed to communicate with the Real World, and that the average person cannot recall any memories of previous lives so that it becomes ever easier to lose oneself in the desire for material wealth.

In addition, the Guiding Spirits of Light could come down to Earth only at set intervals, an additional factor that led to the unforeseen and unwanted side-effect of religious wars and even strife within various branches of a single religion. Satan and the other demonic spirits took advantage of this to slip into the minds of religious leaders to pervert their faith, leading to yet further confusion in the world.

Such a background lends real urgency to our mission to spread the Truth. It also made it necessary to revise the whole system of guardian spirits. Generally, a guardian spirit is either one of the spirits created by the duplicating of a single soul or else it belongs to one of the groups of six made up of a core spirit and five branch spirits. However, it was decided that when a soul comes down to Earth with an important task to achieve, it would be provided with a guiding spirit who is a specialist in the most important facet of the person's life.

This is particularly true of religious leaders, who are assigned a spirit from a level higher than the one they them-selves come from—either from the Nyorai Realm or Bosatsu Realm. In this way, the system of assigning guardian and guid-ing spirits became firmly established, but the disposition of the people living upon the Earth is still very much at the mercy of the various demonic spirits.

6. The Evolution of the Soul

Although the Earth has been plunged into chaos for the past 100 million years due to the influence of hell, it does not mean that the terrestrial spirit group as a whole has regressed. In fact, looking at it over the long term, it has achieved quite remarkable progress, specifically in terms of the evolution of the soul.

There are some souls who were created for the first time on Earth and yet have managed to evolve at a tremendous rate. There are some particularly accomplished souls who have advanced through the dimensions with each successive incarnation, from the fourth to the fifth, from the fifth to the sixth, and from the sixth to the seventh, until they have equaled the high spirits who came originally from other planets.

Although none of the spirits created on Earth has yet reached the ninth dimension, there are some that have progressed as far as the Nyorai Realm in the eighth dimension— to the delight of the grand spirits of the higher realms. This development is exactly what the higher spirits were aiming for when they created the terrestrial spirit group. When they came from their original stars, their hearts were filled with a desire to create a world that was more harmonious, and more advanced than the one they had come from.

Why do souls evolve in this way, and how do they do it? In order to answer the first question, we need to go back to the underlying question: Why did the Buddha consciousness create souls on numerous different levels? This is very important. It does not make sense to say that souls were created with the single object of evolving. Because Buddha is the most highly evolved existence in the universe, there could be no need to produce evolution for its own sake. He did not create souls in various stages of enlightenment for the sake of evolution, but because He expected some good to come out of the process.

To cite a parallel, a parent is a fully developed being, so why would he or she go to the trouble of creating and raising children? Not because the parent simply expects the child in turn to grow into a fully developed parent, but because the parent finds joy in raising the child. The household becomes more pleasurable, and is filled with happiness.

The reason Primordial Buddha created souls on various levels and encouraged them to evolve and develop is not for the sake of development alone but because the process of evolution creates joy. The universe and the various forms of life were created because their evolution is an expression of Buddha's own joy, and a source of general happiness. This is the absolutely basic concept behind evolution in the Grand Cosmos. Buddha watches with love in His eyes over the consciousnesses and souls He has made, caring for them as they strive to evolve and develop to become one with Him.

But how do souls evolve? One way to recognize the stage of development which a soul has attained is by the quantity of light it emits. Looking at the amount of light radiated by a soul in the Real World, its spiritual progress can be calculated at a glance. The same is true of people on Earth. As they progress spiritually and reach a higher level of enlightenment, the light increases and their aura begins to glow. Someone who is spiritually sensitive finds it easy to judge their level of enlightenment.

It is also easy to tell if someone is in contact with hell, for their aura is faint and fuzzy, containing whitish patches moving inside it that are the shadows of the creatures who are possessing them. People in contact conversely with the fourth-dimensional Posthumous Realm (*Astral Realm*) have an aura that encloses the whole body and the back of the head to a depth of between one and two centimeters. Those attuned to the Realm of the Good in the fifth dimension have an aura some three to four centimeters in depth spreading from the back of the head. Those in harmony with the Light Realm of the sixth dimension have a circular aura (a halo) about ten centimeters in diameter. When they reach the upper levels of the sixth dimension and become either Arakan (*Arhat* in Sanskrit) or spirits with a special mission, their aura resembles a small, circular tray that shines with a golden light. Those who are in contact with the Bosatsu Realm of the seventh dimension have a golden, circular aura forty to fifty centimeters wide that spreads from their shoulders. Those in contact with the eighth-dimensional Nyorai Realm are surrounded by an aura one to two meters in depth, bright enough to illuminate their immediate surroundings.

So the degree of development of a soul can be gauged by the quantity of light. In other words, in order to evolve, the soul must become a vessel capable of holding as much of Buddha's light as possible. For this reason, it is important not to let anything into your soul that might deflect away the light of Buddha, and to make a continuous effort, through study, to enlarge the size of your vessel.

7. The Relationship between the Mind and the Soul

I have already used the terms "consciousness," "spirit" and "soul," and although their usage is not always precise, in general tracing the progression from consciousness to spirit to soul is to apprehend the increasing presence of the human element. But are the mind and the soul the same thing?

The mind is the central core of the soul. Just as that bodily organ of the heart acts as the center of the physical body, the mind acts as the center of the soul. It is not to be found anywhere inside the head or the brain cells—as is proved by the fact that when we die and return to the other world, the memories of our lives remain intact. On the death of the body, the brain also dies (and is destroyed, quickly or slowly). But the soul can think, feel, and remember without its physical counterpart. The brain simply functions as a kind of filing cabinet, an information control center; if that control center is damaged, the human becomes unable to make rational decisions or act normally.

Take, for example, a person who becomes mentally impaired due to an injury to the brain. His family may be convinced that he can no longer understand anything said to him, but in reality this is rarely the case. Although he may be suffering from a mental disorder, he understands everything he hears because his mind—the center of his soul—is unimpaired. But although he does understand what is said to him, he cannot communicate this fact to those around him. Although he may not be able to control his body and his

speech under such conditions, as soon as he returns to the Real
World he can function in exactly the same way as anyone else.

Let us imagine the soul as having a shape similar to the
human body. The core of the soul is situated around the chest,
and is responsible for controlling intention, feelings, and
instinct. Another branch of the mind is located in the head, and
functions to control intellect and reason and to transmit com-
mands all over the soul. The part that controls spiritual wisdom
runs right through the body from the stomach, the heart, and the
brain, to connect with its brother and sister souls in the spiritual world. In origin, spirits are a type of energy without form—
but when they live within a human body they create a soul that
is human in shape, with the mind at its center, and it is in this
form that they undertake spiritual refinement on Earth.

Many people like to deny the existence of the spirit or the
soul while they are living on Earth, but very few insist that the
mind does not exist. Of course, there are some experientialists
who claim that the mind is no more than a function of the
brain, yet who despite their vaunted logic still weep when they
are sad, without any conscious decision of the mind to do so.

When we feel sad, sorrow wells up in our chests without
our having time to consider the matter, and tears begin to flow.
When we meet someone dear to us whom we have not seen for
some time, it is only natural to feel an instant warmth and to
want to hug the person. This is not a matter of logic, coldly
controlled by the brain; it is an intuitive reaction of the soul.
People who assert that all conscious thought originates in the

brain are merely quoting materialist dogma that can easily be refuted. There is a good deal more to be said on the subject of the mystery of the mind.

8. The Power of the Mind

It is time to take another look, in greater detail, at what we already know. I explained that human beings consist of a consciousness, a spirit and a soul, all of which were created through the will of the Buddha consciousness, and that the central core of the soul is the mind. I would now like to look particularly at how the mind works and what it can do.

Our feelings frequently communicate themselves somehow to others. When we like a person, our pleasure may be transmitted to him or her, and often he or she at once feels amicably disposed toward us too. The opposite is also true; a person to whom we instantly take a dislike may well somehow sense it and behave toward us in a cold, distant fashion. How does this form of telepathy occur? Let us think about it now.

This power of the mind is actually a power of creation given to us by Buddha. Buddha created the universe through thought—He created three-dimensional space, human souls, and human bodies entirely through the power of thought. Every human being is a fragment of the divine consciousness, a complete microcosm in itself. Accordingly, the power of the human mind is the same and has the same origins as Buddha's own power to create. Our every thought represents the creation of something both here in the three-dimensional world and also in the multi-dimensional world. It is the combined power of the thoughts of humankind that corresponds to the power to create the Real World.

But thoughts are by no means all the same; there are various types and levels of thinking. First, there are the ordinary everyday sort of thoughts that are part of our daily life and experience. A second type of thought comprises what might be called the "concept"; it is a more concrete form of thinking than the first type. Whereas the first type might be likened to overlapping waves endlessly pushing against a flat shoreline, the second type is more consistent and structured; it has a story to tell, and can be visualized and expressed like a movie. In this way, the second type of thought can be likened to waters of a river, which has continuity and direction.

A third and final type of thinking involves "will." This type of thought has definite creative power, a power that is moreover an actual physical force which, when manifested, is known scientifically as psycho kinesis (PK). In the fourth dimension and above, it has a formative power similar to that of Buddha and can be used to create all sorts of things. Even in the third dimension, it is a mental power that has considerable physical force. If you wished to lead people in a positive direction, for example, this "will" could be focused on them to create a sudden change in their minds or cause their lives to change for the better. If, on the other hand, you were to use this power toward people you dislike, they might fall sick, and their lives might change for the worse; they might even die.

This is merely at the level of the individual, but the same is true for large groups. If tens of thousands, or millions, of people all had a desire to create a Land of Buddha or utopia on Earth, and if they all focused their will on the task, a beam of light would spring out of the Earth and this light would filter into the minds of others until the whole world became a happier place. Were this ever to happen, our world would have become the Bosatsu Realm.

Of course, the opposite is also true. If it ever came about that there were enough people on Earth filled with evil thoughts, with hatred, anger, and selfishness, a detached observer looking at the world from a spiritual plane would see negative energy gathering like black storm clouds in various parts and spreading like a physical force to create ever greater chaos. Evidently, then, the human mind has the power to work miracles for good yet at the same time it can be an awesome force of evil. It is incumbent upon us all, therefore, to be very careful in how and what we think, to retain control over the power of our mind.

9. One Thought Leads to Three Thousand Worlds

I now want to amplify what I have said about the power of thought, and in particular to mention the precept that one thought leads to three thousand worlds,[1] a proposition that was often quoted by the Chinese Buddhist philosopher T'ien-t'ai Chih-i (538–597). His preaching undoubtedly followed the gist of a message addressed to him directly by the consciousness of Shakyamuni Buddha, expressed in something like the following terms:

All people have the compass of the will in their minds, but the needle swings wildly, never knowing rest. Even priests who have devoted their lives to meditation do not know peace. If a beautiful woman walks by, the needle in their minds swings wildly. If they are confronted by a wonderfully appetizing meal, it moves again. If they see a rival progressing faster than they are in their studies, it veers once more. If they are scolded by their master, it swivels again. In this way their minds can know no peace.

The true aim of enlightenment, however, is to achieve harmony—and that is impossible if the mind is in constant turmoil. T'ien-t'ai Chih-i, you must study and spread the message. People will never know peace as long as the compass-needle in their minds remains in motion. In the same way that the needle of a compass points north, so should the needle of the mind point in the direction of Buddha. As the Pole Star hangs in the night sky to guide people in the physical world, you must guide them spiritually, T'ien-t'ai Chih-i, and teach them that they must live in the mind of Buddha.

The mind is truly mysterious. If your will is filled with disharmony, your mind will be open to the Ashura

*Realm, the Hell of Strife, and you will know nothing but
strife and destruction. If your mind becomes fixated on
sexual passion, it will connect to the Hell of Lust and
allow the dead to take over your mind. As a result, you
will become the instrument of the dead, craving after
men or women simply to satisfy the lusts of creatures
who have already died. Even those who search for religious awakening may have the compass-needle of their
minds deflected, becoming proud and boastful, teaching false "truths," believing that the voice they hear is
that of the Nyorai and Bosatsu and never suspecting
that it is in fact the voice of the Devil. Poor, deluded
searchers for truth as they may be, they mislead people
and in due course plummet into the abyss of hell.*

*On the other hand, if your will is bent on good, the
compass-needle of your mind will lead the way to the
Realm of the Good in the fifth dimension, and your
friends and ancestors above in heaven will smile on
you. People who try always to save others and who
have cast aside pride and arrogance, are already in
contact with the Bosatsu Realm and have attained the
state of Bosatsu while still living in this world. Others
whose minds are directed toward spreading the Truth,
whose teachings are reliable, whose character is
impeccable, and whose life constitutes an example for
others, are already in contact with the Realm of the
Nyorai, and the Nyorai in heaven are guiding them.*

Heaven and hell are not merely something for us to expect after death; they are here and now. They are in our minds. The compass-needle of people's minds points up and down, to heaven and to hell—and it is in that context that I say "One thought leads to three thousand worlds." Once people have learned this truth they will stop and think, look back over their lives, look back over every single day, and correct the workings of their minds.

The Noble Eightfold Path was based on this law of the mind that I have now revealed to you. Because heaven and hell exist in the minds of people living in this world, the way they live here will dictate the way they return to the Real World, and this is why they must use the Eightfold Path as a basis on which to live their lives.

The Eightfold Path comprises the paths of: Right View, Right Thought, Right Speech, Right Action, Right Livelihood, Right Effort, Right Mindfulness, and Right Concentration. It is only once these various paths have been mastered that people can remain pure in heart and reach fulfillment as human beings.

Use the Eightfold Path to align your thoughts and deeds before spreading the truth that "One thought leads to three thousand world." This will lead you to enlightenment, and to the enlightenment of all humankind.

10. The Noble Eightfold Path: An Interpretation for Today

People are blind. For as long as we live in this world we have only our five senses to guide us and are quite unaware of the worlds that lie beyond. But the real meaning of life exists outside what we can perceive through the five senses. Although it may seem paradoxical, we can nonetheless use our five senses to obtain a clue to understanding what lies beyond them. We should not be content simply to bewail our blindness but should strive to sharpen our senses to allow us to grope toward the truth of life. It is through such efforts that the Eightfold Path may be revealed.

The Eightfold Path is the way to human perfection. It is the knowledge that shows us how to turn off the road that leads us astray and live a proper life. There is no perfect guidance on how we should live our lives, for we are faced in life with a series of problems that vary according to our environment, experience, knowledge, and habits. No one else can solve these problems for us; we are each responsible for doing this ourselves. If we stray from the correct path, it is our fault, and we may or may not find someone to put us back on the right course.

For this reason, we must spare no effort to discover what is meant by "right," in relation to the direction of our lives. What on earth should we use as the standard of "rightness," then? To answer this is the task of a true religious leader—it is the object of my own mission in this world. To know "rightness" is to know the mind of Buddha. It is to make a science of the life of Buddha. It is Buddha's mind that decides what is good and what is evil, what is true and what is false, what is beautiful and what is loathsome. To learn Buddha's mind is to study the nature of the energy of the light of Buddha. In other words, to achieve an understanding of Buddha one must exert oneself to the full.

This is the absolute, genuine Truth. The standard of "rightness" outlined below is an invaluable guide to the way you should live your daily lives, developed from the paths that make up the Eightfold Path. But even before you start, a prerequisite for following the Eightfold Path is faith in the Truth.

Right View

Did you see things rightly, the way they are, based on right faith? Did you observe people rightly? Did you interact with people with a compassionate heart, a heart like Buddha's? See if you have trouble accepting right perspectives on the world and on life.

Right Thought

Did you think rightly? Are your objectives of spiritual refinement correct? See if you have any negative thoughts in your mind, such as greed, anger or complaints. See if you have ill thoughts that cause others harm. Check whether you have become arrogant or if you have doubted Buddha's Truth. See if you have thoughts that go against Buddha's Truth. Did you make right decisions?

Right Speech

Did you speak rightly? Check whether you said anything that you feel guilty about. See if you have hurt others with words you used. See if you have lied about the level of your enlightenment. See if you have misled others by flattering them. See whether you said things that caused distrust between people and led to discord.

Right Action

Did you act in the right way? Check whether you have failed to observe the precepts required of seekers. Check whether your hands, your legs or other parts of your physical body have committed crimes such as murder, violence, or theft. See if you have become involved in sexually immoral behavior such as extramarital affairs, obscenity, prostitution or an obsession with pornography. Do you respect all living things? Check whether you made offerings to the Three Jewels—the Buddha, the Dharma, and the Sangha.

Right Livelihood

Are you living rightly, are your actions, speech, and thoughts in harmony? Check to see that you are not living obsessed with alcohol, tobacco, gambling or drugs. Check whether you are feeling dissatisfied with your daily life. Are you learning to be content? Do you feel grateful for all things? Have you used each twenty-four hours given to you by Buddha properly?

Right Effort

Are you studying Buddha's Truth in the right way? Check to see if your aspiration for spiritual refinement remains strong. Did you part from evil and sow seeds of goodness? Check whether you are neglecting to make efforts in the right way.

Right Mindfulness

Are you able to calm your mind and have right life plans in relation to your spiritual refinement and your contribution to the creation of an ideal world, or utopia on Earth? Are your prayers for self-realization in accordance with Buddha's mind, are they contributing to heightening the level of your enlightenment, and refining your character? Do you understand Buddha's Truth deeply and do you remember the teachings correctly?

Right Concentration

Are you taking time for mediation regularly? Do you reflect on the mistakes you have committed in the past, look back on your day and express gratitude to your guardian and guiding spirits before you go to sleep? Are you in a habit of knowing peace of mind through the practice of meditation?

Following the Eightfold Path leads to freedom. It is an ancient path that has not lost any of its relevance and still today teaches us the "right" way to live. The life of one who travels along it is one of progress and advancement both in this world and toward the ultimate state of Buddhahood.

Notes

1. One thought leads to three thousand worlds: The Tendai sect of Buddhism preaches that everything that exists falls into one of ten categories: 1) an outward appearance; 2) an internal character; 3) the body which defines an individual life; 4) a potentiality; 5) an external activity; 6) the immediate cause of events; 7) causes of events secondary to but related to the immediate cause; 8) the resulting events or outcome of the immediate and secondary causes; 9) the benefit of the outcome; 10) the interrelation between factors one to nine.

On top of this comes the teaching of the Interrelated Ten Worlds (or Realms). The ten worlds are: Hell, the Realm of the Starving, the Realm of Beasts, the Realm of Conflict, the Realm of Humankind, the Heavenly Realm, the Realm of Those Who Hear the Word of Buddha, the Realm of Those Who Have Awakened to the Truth, the Realm of Bosatsu, and the Realm of Buddha. Each of these Realms contains within itself all the other nine Realms.

The corollary of this is that because there are ten realms in the other world and ten levels of mind in this world, overall ten types of people may each have one of ten types of mind—so that altogether there are effectively a hundred worlds and a hundred types of mind. Superimposed upon the hundred worlds are the ten categories listed above, which by multiplication gives a total of a thousand worlds. Superimposed in turn on this total, and in relation to the human experience, there are three further kinds of world: 1) the world of relationship with others; 2) the world of the individual, to which the five elements of form, perception, feelings, will and consciousness all contribute; and 3) society, the community, or the state. These, then, by multiplication take the total up to three thousand worlds. (In this sense the word "world" refers to a world that is distinct in time and space.)

This can only be a brief summary of the theory behind the concept "One Thought Leads to Three Thousand Worlds." For a more detailed explanation, see "The Essential Doctrine of the Lotus Sutra" or "Discourse on Meditation and Contemplation" by T'ien-t'ai Chih-i.

The philosophy is very Chinese in character, but it essentially states that because there are as many as three thousand kinds of mind, there is almost no limit to the number and nature of thoughts that can arise in human consciousness.

Three: The River of Love

1. What Is Love?

Let us consider the meaning of the word *love*. It is surely true to say that love is a most precious thing, a subject of primary concern to us all. It is undoubtedly the most important, the most radiant thing we experience in the course of our lives. People are fascinated by the very sound of the word: *love*. It is full of hope. It is full of passion. It is full of romance. If today was the last day of your life and you knew you were to die in the evening, you would still be able to set off on your journey into the next world with a smile on your face as long as you knew that at least one person loved you.

A life without love could be compared to the journey of an exhausted man through an endless desert. Love, on the other hand, creates oases filled with flowers in full bloom dotted here and there along the desert road of life. But what exactly is love? Who has ever been able to communicate the full measure of what love is? A writer? A poet? A philosopher? Is it only truly religious people who fully comprehend this in the last analysis? How positively can we understand love? How profoundly can we gauge the true quality of love? This is one of the challenges offered to humans—and one of the problems. It is one of the joys, one of the blessings, as well as one of the afflictions, and one of the torments.

Love is both extremes. True love brings the greatest happiness possible, whereas false love brings the greatest unhappiness. Love provides the most joy in this life, but when it is misunderstood it brings the most misery.

We must try to understand the true nature of love in order to achieve the greatest happiness for ourselves. When we finally become capable of this, we shall see a single beam of light shining ahead of us and God waiting for us at the end of it, a smile on His face, His arms spread in welcome.

Jesus is the master of love, a specialist on love, and is himself the embodiment of God's great love. Modern humankind is obliged to seek out the meaning of true love because love has rarely been so misunderstood as it is now. Few periods in history have ever elapsed in which humanity has turned its back on love to the extent that it has today. Indeed, only briefly toward the end of Atlantis, or in Sodom and Gomorrah of the Old Testament, do we find a similar situation.

Because love is so little in evidence and so little understood; it is a topic on which people are always asking questions. I mean to approach the subject head-on, and to answer the questions as best as I can. In the meantime, the best possible context in which to survey love from the viewpoint of life, the world, and a search for Truth, is undoubtedly in combination with a study of the Eightfold Path. The Path and love together probably present the ideal form of spiritual practice in the modern world, one that might even be referred to as the gospel for modern humanity.

2. The Existence of Love

Despite the multiplicity of opportunities that people have to think about love, it remains something that no one ever actually sees. Love is not something that we can take in our hands and examine, nor can we bring it out to show people. Yet it is something that we know exists; it is something that we all believe in. It is this intense desire to track down love, a love we can be sure of that sends us on our endless journey.

If love is something that no one has ever seen, something that no one has ever touched, does this mean it is nothing more than a fantasy—a mirage? No, because if we think about it, we realize there are any number of things that we cannot see with our eyes or touch with our hands but that we authoritatively believe to exist. Take the wind, for example. You cannot see the wind with your eyes but you can watch the leaves blowing in the air, or hear the wind sigh in the trees, and this is proof positive to you of its existence. You can feel it on your skin, so you know its gentleness, its coolness, its strength—and you describe this touch as being "the wind." But you cannot shut it up in a box and bring it out at a later date to show someone else.

In this respect love is very similar to the wind. Everyone believes in it and can feel that it exists, but no one can prove it objectively. In this way too, the existence of love bears a remarkable similarity to the existence of God.[1] Many people have spoken of God and many people believe in His existence, but no one has been able to produce Him and show Him to us. Over the ages a vast number of very worthy individuals have spoken of God in terms of religion, philosophy, poetry, and literature, but none of them has been able to offer any real proof of His existence. Even Jesus Christ was unable to show God to the people; he could not point his Father out to people and say, "That is God."

Christ said that those who heard him speak heard the words of his Father in heaven, that his Father's spirit had entered into him and was speaking through him. In the same way, the things he did were the works of his Father; God was acting through him.

What he was telling people was that they could experience God through what he said and what he did. And, indeed, people felt God when they listened to Jesus' authoritative words and were persuaded by them into changing their lives.

It is a fact that the things people consider most important in life are things it is impossible to prove exist. No matter how times change, that seems to remain the situation. Throughout history, the things that people have held in the highest esteem have been:

God, love,
bravery, wisdom,
goodness, kindness,
beauty, harmony,
progress, compassion,
truth, sincerity,
selflessness.

The universe is filled with all of these qualities—in fact, there is not one inhabitant of the Land of Light who is unfamiliar with them. Here on Earth, however, it is difficult to prove conclusively that they really exist. The reason for this is that all of these important qualities exist in the Real World in the fourth dimension and beyond, and it is difficult to prove their existence in terms of three-dimensional standards and symbols.

The supreme being I regard as the Primordial Buddha (Almighty and Everlasting God) exists on the ultimate plane beyond the twentieth dimension. It is utterly impossible to prove His existence using the references of the third dimension. This is why we have to have faith. The dictionary tells us that *faith* is a belief and trust in God. Belief involves acknowledgement and acceptance; trust involves reverential acquiescence.

Jesus said, "God is love,"[2] and, although love is undoubtedly one of the attributes of God, it can be argued that Jesus was actually saying more than just that. His philosophy on the subject of love nearly 2,000 years ago in Nazareth may perhaps be expressed as a reconstruction of what might have been his own words:

It is impossible to prove the existence of God. The nearest I can manage is to compare God's existence with that of a similar power, of love. No one can prove the existence of love, but despite its being impossible to prove, everyone knows its wonder. They know the

beauty of love, and struggle to attain it for themselves.
They believe in the power of love.

Faith in God is very similar. Anyone who believes
in love must believe in God. Anyone who believes in
the power of love must believe in the power of God.
This is because God is love. Look at me, Jesus Christ,
the Son of God, as I work through love. My actions are
not my own, but those of my Father in heaven who
works through me. If you want to see love, look at what
I do. In my actions there is love. In them there is God.

3. The Power of Love

As far as I know, love is the strongest force that exists on
Earth. Even outside this world, in the Real World and in many
worlds in other dimensions, love remains the strongest force.
In fact, as we climb the ladder between the dimensions, we
find that love grows steadily stronger. The reason for this is
that love is a power of *joining*, of uniting and strengthening.
Powers of rejection weaken each other. Powers of joining, on
the other hand, double or treble the power of the individual.
That is why love has no enemies and no enemies try to stand
in the way of love.

Love is like a military tank. It climbs hills, descends val-
leys, fords rivers, and crosses swamps. It is not halted by the
fortresses of evil but continues on regardless, like a tank.

Love is light. Love shines through the dark night, it shines
through the past, it shines through the present, and it shines

through the future. It shines in heaven, it shines on Earth, and it shines in our hearts. It easily melts away all the evil in this world, and pours out its warmth on all sadness.

Love is life. All the people of the world live on the bread of love; they live by love's strength and think of love as the flame of life. In other words, love is everything. Without love there is no life; without love there is no death; without love there is no path; without love there is no hope. Love is the be-all and end-all. It is bread, it is life.

Love is passion. Passion is the power of youth and the belief in boundless possibility. In that fiery energy is truth, and the ceaseless throbbing pulse of life.

Love is courage. Without love, people would not be roused to action. Without love, people would not be able to stand against death. Love is the flame that lights the fuse of truth; it is the arrow dispatched against delusion.

Love is a communal vow. People live, talk, and walk together all in the name of love. Without the bonds of love, people would wander in their separate ways and simply wait for dusk to fall.

Love is good words, good memories, good surroundings, and good music. God creates the world through words, and love creates people through words.

Love is harmony. It is through love that people grow close to each other, forgive each other, nurture each other, and create a beautiful world. Within the circle of love there is no anger, no envy, no jealousy. Within the circle of love there is only a great harmony in which everyone looks after and cares for each other.

Love is joy. Without love we would not know true joy; without love we would not know true happiness. Love is the expression of God's joy and is the power that can banish sadness from this world. Love is joy, and that joy brings forth further love, which in turn produces yet more joy.

Love is progress. A single element of love makes for a single element of progress; a single spark of love creates a single spark of light. Days filled with love are days filled with progress, days filled with light. For God waits at love's destination. Countless holy spirits wait at love's destination. Where there is love there is no retreat. Where there is love there is no fear. In love there is only progress. In love there is only improvement. Love is simply the act of flying to God's side.

Love is eternal. Love exists in the past, the present, and the future. There has never been a time when love did not exist; there has been no time during which people have lived without love. Love is like a shining golden wing that cuts through all ages. It is Pegasus flying through the sky. Love is living proof of eternity; love is the hunter who catches the eternal now.

Love, finally, is prayer. Without love there can be no prayer, and without prayer there can be no love. Through prayer, love becomes a more positive force. Through prayer, love can achieve everything. Prayer strengthens the power of love and is the secret power that deepens love. Through praying to God, love can be achieved; and through praying to God, love can be realized.

God is love. Love is God. It is through the power of prayer that love becomes God. In this way, people can realize the full power of love.

4. The Mystery of Love

Love is a true mystery. Its depth and height are infinite and impossible to measure. The more we think about love, the greater its substance, the more profound its experience. God refrains from revealing Himself to us but sends "love" down to Earth in His stead. By making us learn from this, not only does He allow us to understand His true nature but He presents us with a subject for study.

Through love we can feel the power of a force that is invisible to human eyes. This is what makes it a mystery.

Here is a short story about the mystery of love:

Once upon a time, there lived an old man. Although he was very advanced in years, he did not have any children or grandchildren and so he lived a very lonely life. His home was a small shrine on the edge of the village, where the local children would often come to play. The most mischievous of the children was a boy of thirteen called Taro, who had lost both his parents and was being brought up by his married sister.

One day, Taro was playing on the steps of the

shrine when three sparrows settled on the step next to where he was sitting, and one of them started to speak.

"The sun is the greatest thing in this world," it said. "It is thanks to the sun, shining in the sky, that our world is filled with color and the trees, flowers and grasses can flourish, that the corn is bowed with the weight of its seed and we sparrows can feast at will.

"If the sun were hidden, the world would become pitch black and nothing could live. It is because we sparrows always give thanks to the sun that we remain humble and reverent, and do not kill others of our own kind. But humans take it for granted that the sun continues to smile down upon them, and they are filled with pride and do whatever they want. They fight and swear at each other, and some of the crazier ones even start wars. If they go on like this, the sun will have had enough of them one day and hide himself away."

Hearing this, another sparrow said, "No, the greatest thing in the world is water. If it were not for water, no living thing could exist. Without water even the trees and grasses would die before the week was out. Without water there would be no corn or rice and we would all die. Without water none of the other animals could live for more than a week. For this reason, water can be called the seed of life—and I believe it is the most wonderful thing in the whole wide world.

"It is because we sparrows give thanks for water

*that we are able to live in harmony, but humans think
it is free and do not pay it any respect, preferring to
waste their sweat working to buy useless jewels and
necklaces. We sparrows are quite satisfied looking the
way that God made us, but humans spend all their time
trying to think up ways to make themselves look more
beautiful. They all want to be more famous than their
neighbors, to be richer, to be more beautiful, when
really they are all just fools."*

*Finally the last sparrow reluctantly began to
speak.*

*"What you say is true: the sun and water are mar-
velous gifts. But the most valuable thing in the world
is something that most people don't even think about,
something whose bounty they do not even notice. No
one takes any thought for its existence—but I think the
most important thing in the world is the air we
breathe. If the sun were to hide its face or the water to
dry up we would still be able to live for a few days, but
without air we would be dead in minutes. You might
realize this now that I have alerted you to the idea, but
most of the time people do not pause to realize just
how valuable air is.*

*"When we fly through the sky, we take a lungful of
air and are thankful for it. Even the fish, when they are
in pain, poke their heads out of the water to breathe
the air, and give thanks for it. But human beings in*

their pride think that it is their own intelligence that allows them to build aircraft that can fly through the air. They are wrong; it is the air that allows their machines to fly. Air does not make any demands of humans or birds when either of us wants to fly through the sky. We give thanks to the air, but I have never noticed a human doing it."

Listening to the three sparrows, Taro became very sad and fell deep into thought. He had been brought up to believe that humanity was the lord of creation, and had never heard anyone speak as the three sparrows had done. It was all too true; he had never felt gratitude for the sun, for water, or for the air, and he realized what unthinking fools humans were. Surely they must be worse than the sparrows.

He jumped to his feet and ran up the steps toward the shrine, scaring away the sparrows as he went. He told the old man who lived at the shrine the story that he had heard from the sparrows, and afterwards burst into tears, saying that if humans were such fools, he wished he had been born a sparrow instead.

The old man smiled gently and replied, "I congratulate you on what you have learned. You are right. Humans are all fools and have lost sight of the most important things in life. However, foolish though they may be, the ugliness of their sins can be forgiven as long as they love each other. Humans may contain ugli-

ness in this way, but the way to get rid of the ugliness is not by concentrating on it, for no matter how hard you concentrate on the ugliness, it will not disappear.

"In order to forgive humanity of their sins and expel the ugliness, God gave us love. It is because of this love, through its mysterious power, that He allows humans to remain the lords of creation."

5. Love the Invincible

Love is the greatest power in the universe; love is invincible. People are faced with various hardships during the course of their lives, and it is through these hardships that the soul learns. It has been planned this way since the beginning. What are the hardships I am talking about? I mean sickness, destitution, frustration, a broken heart, the failure of a business, the loss of friends, separation from loved ones, a meeting with someone we loathe. On top of these difficulties we grow older, become ugly, lose our faculties, and eventually die.

If we look only at these problems, our lives might seem to be filled with sorrow and distress—but this distress is not without meaning, our sadness is not without its purpose. Such sorrow and distress provide us with a choice. They offer us the choice of living a life of giving or living a life of taking.

The basic property of love is that it involves giving. To love is not to take God's gift and keep it to oneself, but to share it with others. God's love is infinite, so no matter how much we may give to other people it will never run out. God keeps us supplied with love. I must emphasize the point: love is above all a matter of giving—let no one misunderstand this.

Readers who are at this moment in pain and distress because of love, may I ask you why you are suffering? Exactly what is the reason for your pain and distress? Is it because you have given love? Or can it be that you have been expecting something in return? True love expects nothing in return. True love is a love that gives, an act of just giving unconditionally. The love that you give does not originally belong to you; it comes to you from God. In order to return this love to God, it is necessary for us to love other people.

Does your pain and distress stem from the fact that although you love a person, you feel that your love is not being returned, that the person does not love you? It is when we feel that people do not love us as much as we think they should that the opportunity provided by love causes us suffering. But we should not look to other people for a return on our love, but to God.

But what is it that God gives us in return for our love? The more we love other people, the closer we, as people, become to God. That is what He offers us. Let us look at what we know about God. God is like the light of the sun, pouring infinite love and grace on all living creatures without asking anything in return. The life of each one of us embodies an energy that God has provided without demanding a price.

We should, therefore, make a start by giving. We should live our lives day by day thinking how we may afford happiness to others through love. By giving, we cast the light of love on the hearts of those who are lost. By giving, we can save people from distress and frustration, and instead provide wisdom and courage.

Give wisely. The simple donation of material goods can hardly be considered real giving. If it simply satisfies our desire to appear charitable, it is not giving at all. True giving is the path to true caring and fostering, and wisdom is essential to proper nurturing of this kind. That is why we must live a life of giving with wisdom and with courage. We must be prepared to give love without expecting anything in return.

Love knows no enemies. Love is invincible. Love is like a vast river, a river that flows from an infinitely wide source to an infinitely distant end. No one can fight against the power of the river. It gives all and has the power to wash everything away. No evil exists in this world that has the strength to fight forever against the power of love.

6. Stages of Love

True love is a love that gives, a love that expects no reward. But love itself develops through several stages, a process that few people think about.

The first step in the development of love is *fundamental love*: love at the personal, family, and ordinary social level. This is perhaps the most easily understood type of love. It is the love of a parent for a child, of the child for its parents, of a man for a woman, of a woman for a man, the love between friends, or even the love between neighbors. In a wider meaning, love for society or the community as a whole could also be included in this "fundamental love."

Of course, this love is a love that gives. The basis of fundamental love is that a person has an interest in the object of his or her love. The goodwill that a person shows toward the object of his or her interest is a manifestation of fundamental love. This love is the most basic and most common form, but in practice it often proves to be a love that represents a source of difficulty.

It is no exaggeration to say that the world of the third dimension would become heaven if it could be filled with fundamental love. Everyone is born with an innate understanding of the beauty of fundamental love, for it is a love we can all expect to experience. Indeed, it is part of the way that people are made that they receive pleasure from giving love. The problem, however is that although it is all very well to have an understanding of love, it is worthless if it is not put to use. If this love could truly be achieved on a planetary scale, this world would be transformed into the Realm of the Good in the fifth dimension while still remaining in the third dimension. This is how fundamental love could constitute the first step to creating a heaven on Earth.

The second stage in the development of love is *spiritually nurturing love.* Anyone and everyone is capable of experiencing and giving fundamental love, although they may or may not wish to do so. But spiritually nurturing love is a love that not everyone is capable of. After all, only mature people are able to foster potential in others; only people whose talent and efforts have enabled them to teach are in a position to nurture others.

This nurturing love is a love that leads and guides, and in order for it to produce results, it must first exist as an inner quality of a character better than mere fundamental love. It is impossible for the blind to lead the blind.

Spiritually nurturing love is an intellectual love, a reasoning love. Only a person with a high intellect can understand the true state of humanity and society; only a person with superior reasoning can understand all the problems, take the necessary steps to resolve them, and truly guide others. People who embody this nurturing love must sometimes use it with the burning passion of the true teacher to save the spiritually degenerate and lead them back to the right path. If they are incapable of this, they cannot truly practice spiritually nurturing love.

According to this definition, spiritually nurturing love could be equated with love in the Light Realm in the sixth dimension of the Real World. Of course, there are spiritual leaders in this world who are capable of exhibiting this love, but in their hearts they are already in contact with the sixth dimension anyway.

So far, we have looked at fundamental love, which entails taking an interest in our fellow humans and wishing them well, and spiritually nurturing love, for which we must strive to improve ourselves in order to guide others. Both of these

are wonderful forms of love, yet neither on its own is sufficient. There is a love that surpasses ability, intelligence, or
hard work, and that is the third stage of love. This is *forgiving
love.*

Those who practice forgiving love should have experienced a sudden uplift to a state of high holiness. This is
because forgiving love transcends good and evil and is
restricted to those who have devoted themselves totally to
their mission. The people in this state have come to the realization that the inhabitants of the materialist third dimension
are blind in the spiritual sense, groping pathetically for what
they cannot understand. To have come to this realization
means that they have also become conscious of their own spiritual emptiness, cast it aside, and by doing so experienced a
religious reawakening. Only people who have discovered the
light through their own suffering are able to see through the
masks of others and to love their divine nature. It is a state that
occurs solely in those who are both magnanimous and generous, whose virtue surpasses intelligence.

Such people see all others as the children of Buddha, as
reflections of God, and even perceive the qualities of Buddha
in those who would be their enemies (very much a form of
transcendental wisdom). The state of forgiving love is the state
of the spiritual master, the Bosatsu, or the Bodhisattva. The
people capable of forgiving love are messengers from the seventh dimension, and their hearts remain in the Real World of
the Bosatsu.

At the same time it should be understood that those who practice forgiving love—that is, the Bosatsu's love—are just as fervently constant in their unforgiving hatred of the Devil and his activities. The Devil in this case corresponds to anyone who obstructs God's love for humankind, anyone whose very being is the antithesis of love. The Bosatsu battles with the Devil using the weapons of faith and selfless wrath. Once the Devil realizes that he will never be able to gain the final victory over Buddha (God), it will be possible for him to pass through the gates of forgiveness. It is because of this that active forgiveness plays a necessary role in forgiving love. But there are even higher states of love.

7. Love Incarnate and God's Love

There is a form of love higher than forgiving love: I call it *love incarnate*. This love is no longer the love of one person for one person or group, and it surpasses even differences of rank. Love incarnate is when a person's very existence—the mere fact that the person's life touches yours, however momentarily—enables you to resolve your doubts to change your whole life, even to experience a religious awakening or enlightenment. The mere fact that such a person lives is enough to bring light to the world and give hope to those who share the same era. It is not love that is consciously of or for anyone; it is not love that uses exquisitely beautiful words; it is not love that is ultimately kind and generous to others; it is a love that corresponds simply to existing, being there, and influencing the world in that way. The character of a person is itself love incarnate. Such people shine out through the history of humankind.

If forgiving love is the love generated by particularly virtuous and holy people, then love incarnate is the love of a great person who holds a high place in human history, who is the contemporary light of the world, the very spirit of his times. Love incarnate is, as I have said, not the love of one person for one person, it is the love of one person for many, for all—a love that radiates out in all directions. In other words, it is another form of light.

This description may have already indicated which dimension this stage of love belongs to. It belongs to the Realm of the Great Masters (Nyorai, or Tathagata) in the eighth dimension. When love incarnate is expressed through a person in a particular historical period, that person is in reality a facet of Nyorai and represents unimaginable compassion for humankind. This compassion is the light of love that shines throughout the world; it is not a selective love that produces different shades depending on whom it touches. Compassion is both absolute love and impartial love.

The highest form of love available to humankind is that of the ninth dimension. This love may be described as *love of God Incarnate,* or *the love of the Savior* (in that it constitutes salvation). However, I do not recommend anyone to seek to attain this sort of love through spiritual discipline, for this love is restricted to those chosen by God (Buddha) as divine instruments, as His supreme representatives.

Furthermore, "the Love of the Savior" or "the Savior's Love" is an expression used by some religious teachers who have no concept at all of the Truth. What awaits them after death is not the paradise of the ninth dimension, but the deepest pit of hell in the fourth dimension. For to preach false teachings as though they were the words of God (Buddha) is the worst crime imaginable, worse by far even than murder or robbery. It even depraves the eternal soul as well as the life of people in this lifetime. So when we consider the stage above love incarnate, we should simply be content to think of it as God's love (Buddha's great compassion), a love that offers divine guidance, a love that offers us hope of evolution.

At the other end of the scale, of course, there is the love stemming from the fourth dimension, which can be referred to as *instinctive love*. Depending on how it is used, this can be used to communicate with both hell and the Astral Realm, but it is not a love that should be aimed for in anyone's spiritual refinement.

To summarize, the stages of love are:

Instinctive Love   (fourth dimension)
Fundamental Love   (fifth dimension)
Spiritually Nurturing Love (sixth dimension)
Forgiving Love   (seventh dimension)
Love Incarnate   (eighth dimension)
God's Love   (ninth dimension: beyond
                            human wisdom)

Love is what we aim to attain through our spiritual refinement, so it is incumbent upon us to make sure we have a good understanding of these various stages.

8. Love and the Noble Eightfold Path

The Noble Eightfold Path was featured in Chapter Two; in this chapter we have looked at the various stages of love. I would now like to consider the relationship between the two.

The Eightfold Path provides directions by which a person may live a life that is "right" and encourages contemplation on the subject that can ultimately lead to enlightenment. Enumerating the various stages in the development of love, I have described four levels of love that may be achieved through spiritual practice: fundamental love, spiritually nurturing love, forgiving love, and love incarnate.

Comparing the two approaches, it is at once apparent that the Eightfold Path concentrates on daily practice and daily enlightenment, whereas the stages in the development of love—although having their origins in everyday life—present a mid- to long-term target. But if we see the Eightfold Path and the stages of love simply as ways to achieve enlightenment, the following is produced:

Right View and Right Speech
    lead to fundamental love.
Right Action and Right Livelihood
    lead to spiritually nurturing love.

Right Thought and Right Effort
   lead to forgiving love.
Right Mindfulness and Right Concentration
   lead to love incarnate.

Let me explain how this works. First, how do Right View and Right Speech lead to fundamental love? Fundamental love is a love for a person in whom you take an interest. In order to show the object of your love a most appropriate form of goodwill, you must begin by perceiving him or her correctly, in the light of "true faith." It is important to be able to distinguish between right and wrong; you must be able to see clearly what the object of your love wants and what is troubling him or her, without allowing yourself to become confused by preconceived ideas. If you are sure that you are seeing clearly, the next thing is to be able to speak in the right way—to be able to offer valuable advice without in any way hurting the feelings of the object of your love. You should use words that will warm the person's heart or help the person when he or she is in distress.

Next, how do Right Action and Right Livelihood lead to a spiritually nurturing love? Right Action involves acting with propriety. In the days of Buddha, it entailed observance of the religious precepts and ensuring that the body did not commit any sin—in other words, people were not to kill any living creature (including humans), to steal, or to have a physical relationship with a person of the opposite sex other than their spouse. Today, it means that one should not use violence, steal, or be unfaithful to one's spouse, and that one should work to become a good member of society. Everyone should act with respect for other people's rights and dispositions, and try to enlighten other people through the betterment of their own personal character as a member of society.

Right Livelihood means that one should perform one's duty and live an upright life. One must avoid any activity that contravenes the precepts of Buddhism and would, therefore, lead straight to one's own destruction. To take it to its logical conclusion, this means that one must not become an urban terrorist or participate in the sex industry. But at a lower level one should also refrain from heavy drinking, from serious gambling (including betting on horses or saloon games), and from the use of narcotics and other drugs (including tobacco), which can damage one's health. Unnecessary debt should also be avoided, for it can hardly be described as Right Livelihood.

We should also remember that we cannot live without other people. Throughout life, we rely on the help of others and must be able to live in harmony with a variety of people. In other words, a life of propriety—that is, a life within true faith—enables us to live in unity and creates a suitable basis on which spiritually nurturing love can prosper. It creates an environment that allows us to guide each other. The more people there are who work to create an idyllic everyday life with their family—that is, the more people there are who enjoy Right Livelihood—the closer the Earth will become to heaven. It is in this way that Right Action and Right Livelihood correspond to the same stage as spiritually nurturing love. ·

The third premise is that Right Thought and Right Effort lead to forgiving love. Let us first consider Right Thought. Right Thought involves thinking correctly, without allowing oneself to be ruled by the Three Poisons (greed, anger, and foolishness) or the Six Worldly Delusions (the Three Poisons plus pride, doubt, and false views), looking upon our relationships with others, and being ready to adjust them if necessary.[3] Do not make the mistake of allowing people's outward appearance alone to be your guide to their true nature—instead try to perceive people as inhabitants of the Real World, and on this basis think about what is the correct way in which to interact with them.

Think back now: Is it possible that you may be wrong about what you believe the people around you to be like? Now think of them as children of Buddha; think of their essential form. You should be guiding each other toward perfect harmony. If you can think correctly, your heart will always be filled with patient tolerance that will in turn enable you to stay benevolent as if you are embracing all. When you can reach and maintain the state of Right Thought, therefore, your heart will naturally also attain the state of forgiving love.

It is the same with Right Effort. To travel the right road, to endeavor to acquire the Truth, to struggle ever onward shunning temptation in an effort to fill one's heart with goodness, is to live one's life in a state of ever more profound enlightenment. By working hard on the path to the highest consciousness that we call Buddha, you can double the virtue you attain. There is no anger on that path, no grievance, no dissent or dissatisfaction, and no envy—only literally right thoughts that allow perfect harmony to spread into and over the world. If the mind remains unshakeable in its purpose, it can produce the power to purify even sinners. So the more we undertake Right Effort, the deeper our religious discernment becomes and the wider the state of forgiving love unfolds before us.

The fourth premise is that Right Mindfulness and Right Concentration lead to love incarnate. Right Mindfulness involves focusing one's life on the Truth—in an atmosphere of calm reflection, planning one's life in accordance with the Truth and praying for self-realization. This is certainly Right Mindfulness, but what exactly is the "self-realization" prayed for by those searching for the Truth? It is the perfect form of a child of Buddha. It is the state of becoming one with Buddha, the state of Nyorai (Great Master). It is the ultimate human form, and those who achieve it are revered by all. The existence alone of such a person in the world provides a light for the human race. This is Right Mindfulness: the ultimate state that religious people can aim for.

Right Concentration also refers to the ultimate state that religious people who seek the Truth can attain. Over the millennia, holy men and women have practiced a great number of different methods of meditation in an effort to achieve enlightenment and commune with the superior spirits in the higher dimensions. Right Concentration involves living in a daily state of self-examination, communing with one's guardian spirit, and then communing with the guiding spirits to carry out Heaven's will. Eventually one may even be able to contact the Guiding Spirits of Light, the inhabitants of the Nyorai Realm.

The mind of a human being is three-thousandfold. If we can achieve the state of enlightenment of Nyorai, it is possible for us to communicate through Right Concentration with the Great Guiding Spirits of that realm. Everyone who enters the eighth dimension of Nyorai receives instruction either directly or indirectly from the Great Guiding Spirits of the higher level of Light. At the very least, everyone who achieves that state is sure to receive the inspiration to fulfill his or her own appointed task.

The first step in achieving the state of love incarnate, then, is to enter Right Concentration, find liberation from the bondage of this world, and accomplish true spiritual concentration.

To recapitulate some of this section: the Noble Eightfold Path can be divided into four stages that make for spiritual progress:

Right View and Right Speech
Right Action and Right Livelihood
Right Thought and Right Effort
Right Mindfulness and Right Concentration.

To study them in pairs like this helps in self-examination. The order in which I have arranged them differs from the order expounded by Buddha, but it presents an effective means of learning for those who are only just beginning spiritual studies.[4]

An alternative method is to try to develop spirituality by progressing, a stage at a time, through the stages of love. This would mean that once you have managed to achieve fundamental love you should then make the attempt to reach spiritually nurturing love, and once spiritually nurturing love has been attained you should strive to reach forgiving love, and finally your goal should be love incarnate.

Remember: Right Action and Right Livelihood, Right Thought and Right Effort, Right Mindfulness and Right Concentration cannot exist without there having been from the start Right View and Right Speech. In the same way, spiritually nurturing love, forgiving love and love incarnate cannot exist unless from the first there has been fundamental love. In both cases the first stage is of crucial importance.

9. The Love of Angels

Now let us turn our attention away from people who benefit from receiving love and who use it to good effect, and instead consider where love comes from: the angels who inhabit the higher realms of the spiritual world.

The beings we refer to as angels inhabit the Light Realm in the sixth dimension or other, even higher realms. They comprise, that is to say, high spirits with a special mission in the upper realms of the sixth dimension, the Bosatsu from the seventh dimension, the Nyorai from the eighth dimension, and the Grand Nyorai from the ninth dimension, who are also known as the Great Guiding Spirits of Light.

All these beings endow the world with love, but seem to manifest it in different ways. The love of the Arakan (Angels of Light, or *Arhat* in Sanskrit) of the upper realms of the sixth dimension, for example, is expressed in three ways—as the love of those who guard and guide people on Earth, as the love of those who save the souls in hell, and as the love of those who teach the inhabitants of the Realm of the Good in the fifth dimension.

The love of the Bosatsu of Light (angels) from the seventh dimension is expressed in four ways. First, there is the love of those who are born on Earth as religious or social leaders to guide humankind. Second is the love of those who offer themselves as assistants to the Great Guiding Spirit. Third is the love of those who as leaders save the souls in hell. Fourth is the love of those who supply light to the Real World. It is through the mediation of the various Bosatsu that the worlds of the sixth dimension and lower receive the light of Buddha.

The love of the Nyorai of Light (great angels) of the eighth dimension is expressed in five ways. First, there is the love of those who appear on Earth every few hundred years to preach a new religion or to revolutionize an existing religion, and to provide the world with new learning. Second is the love of the Nyorai who guide the Bosatsu: each Nyorai is responsible for teaching tens of Bosatsu, and every Bosatsu (thus) has a Nyorai as a teacher. Third is the love of the great commander whose task is to fight Satan and convert him. Fourth is the love of those appointed to the duty of taking a single beam from the prism of Buddha's light (for instance, the light of love) and disseminating it. Fifth is the creative love of the spirit responsible for establishing new civilizations.

All things live within the love of the Grand Nyorai of Light, the Great Guiding Spirits of Light of the ninth dimension, although that love can nonetheless be categorized in six different ways. First, there is the love of the savior who appears on Earth every few thousand years to create a global religion and to purify the world. Second is the love of the teacher who guides the savior from the Real World while he is on Earth. Third is the love of the Grand Spirit who organizes and presides over the evolution of humankind. Fourth is the love contained in Buddha's seven colored rays of light, which correspond to the love of the supplier of individual light to the eighth dimension and dimensions below. Fifth is the love of the grand spirit responsible for preserving order in the Real World, the love that is used to gauge how the people as a whole are thinking and developing. Sixth is the love of the grand spirit ultimately responsible for creating a plan for the Earth consistent with the plan for the universe.

10. The River of Love
This chapter has considered love in many aspects, the love of

humans as well as the love of angels. But the fact that it has so many aspects itself raises the question: What is love that it flows not only through the third dimension but in all the other dimensions too, from the fourth up to the highest realm? We could describe it as a cascading torrent, the tumultuous surging of the vast river of life, which (in a spiritual sense) is a tremendous flood of energy. To watch it flow from the ninth through the eighth, the seventh, the sixth, the fifth and the fourth dimensions to the third is a magnificent sight, an overwhelming spectacle.

Love may be described as a single, vast river; a single flow that surges from its source to its end without slowing; it is the vigor of life that stops for no obstacle in its path. Love has no enemies. Gazing at the majestic River of Love we can clearly see that it is impossible for anything to stand as an enemy against love. Hell? Do you imagine hell as having the sort of power to be able to stand against heaven, against God (Buddha)? Is your conception of the spiritual world that of a single entity divided between heaven and hell? If it is, you could not be more wrong. The vast River of Love originates with God (Buddha), gushes out with tremendous force, and sweeps over everything in its path. Hell lies in the fourth dimension, not far from the end of the River, and although the estuarine saltwater of materialism, lust, confusion, and evil may try to contaminate the River's waters, they cannot block it and are eventually washed away by its mighty flow.

Love is Light. No darkness can triumph over light, and in

just the same way no evil can stand against love. Hell certainly cannot halt the flow of love; hell certainly does not have sufficient power to oppose heaven—it is itself no more than a cancer lurking in the corner of the world that was created by God (Buddha); it represents no more than a few drops of saltwater that have somehow found their way into the pure waters of the River.

The popular conception has long been that heaven and hell are approximately the same size, and that the angels and the Devil have been engaged in continuous conflict. It is not like that at all. Heaven (the "Celestial Kingdom") stretches from the fourth dimension of the spirit world all the way up to the highest sphere. Hell, contrarily, occupies a stagnant recess, an area shaded from the light, in the fourth dimension. Its area is actually large in human terms, of course, capable of supporting a population of several billion. But just as ice cannot exist for long in sunlight, hell too will disappear eventually. Its influence is overestimated on Earth simply because the spiritual plane on which it is based is extremely close to our world, and what happens in one realm has an effect in the other.

What does hell consist of? Hell consists of forms of conceptual energy that in our world are expressed as, for example, envy, jealousy, mindless rage, petulance, complaint, greed, dissatisfaction, the desire for revenge, pessimism, nihilism, deliberate indecisiveness, cowardice, sloth, self-loathing, hatred, lust, arrogance, selfishness, defamatory language, deliberate deception, lies, aggression, depression, vio-

lence, ostentatious elitism, materialism, atheism, isolation-
ism, tyranny, desire for worldly status, desire for money, and
disharmony.

These forms of energy do not have a specific shape or
nature of their own because they are merely negative energy—
the lack of, or the opposite of, positive forms that do have their
own shape and nature. In particular, hatred, jealousy, rage,
petulance, and dissatisfaction are all no more than manifestations of a lack of love. They only exist where love does not.

In the final analysis, then, the spirits of hell do not possess
the power to stand against the light of heaven, and are basical-
ly no more than "beings without love," "beings who want
love." At their core, spirits in hell want to be loved, they want
to be treated kindly by people; they want love, more love, and
yet more love. Pitiful, pathetic creatures who need to be saved,
they are sick—and the sickness from which they suffer is a
deficiency of love.

We have seen that the basis of love is giving, but the
inhabitants of hell know only desire, they are forever yearning
to receive something. They were once people who did not
understand the true meaning of love, who knew only the
"love" that is taking rather than giving, and so they are suffering in hell. But we could get rid of hell once and for all right
now. It would not be difficult. All it would take is for every
person to understand the true meaning of love: Giving. What
should we start by giving? A giving love begins with gratitude;
we must give thanks to Buddha (God) for supplying us with

what we need. In doing this we will surely be moved to do something for this world that Buddha created, something that represents a willingness for repayment. This is the first step in giving love to other beings, and this is how we should begin.

Notes

1. In this sentence my use of the word "God" does not refer to the Supreme Consciousness but to a deity in general. To be honest I had a mind-picture here of the god Hermes. Everywhere else in this chapter, however, the word "God" is used as the Christians use it, and refers to the Lord El Cantare.

2. "God is love" is an expression of the nature of the Lord El Cantare, the "God of love" as recognized by Jesus Christ.

3. The Three Poisons and the Six Worldly Delusions: The three negative mental functions which contaminate the Buddha-nature of humans, and which are known as the Three Poisons, are greed, anger, and foolishness. Together with pride, doubt, and false views they comprise the Six Worldly Delusions which orthodox Buddhist thought considers to represent a major force that prevents people from living "right" lives and sets them on the road to hell. On the other hand, it is often said alternatively that humans are subject to no fewer than 108 worldly delusions—which illustrates the quantity of evil we are confronted with. But the power of Right Thought is infinite.

4. People who are serious about seeking the Truth and who have joined The Institute for Research in Human Happiness (IRH) are recommended to study and practice The Words of Emancipation—Buddha's Teaching: The Eightfold Path, comprised in our basic sutra, The Dharma of the Right Mind.

Four: The Ultimate Enlightenment

1. What Is Enlightenment?

To start with, enlightenment is to know your true self, to know the way the world works, and to know the purpose and mission of life. It is generally something that comes through religion, but in philosophy, too, there is a strong basic urge to achieve this knowledge. The aim of philosophy, after all, is to attain a knowledge of Truth and to reach an intellectual understanding of the mysteries and workings of the world.

It is open to debate whether or not the teachings of Confucius represent a religion, but there can be no doubt that their objective is to lead humankind to perfection through moral fulfillment. It was by teaching morals that Confucius was able to guide people to enlightenment.

Throughout the ages human beings have searched for enlightenment. Even though they may not understand exactly what enlightenment involves, everyone has an innate desire to advance themselves spiritually.

This chapter approaches enlightenment from a religious point of view. Such a viewpoint, of course, embraces the philosophical desire to arrive at the truth and the Confucian pursuit of moral perfection for humankind—but the important difference with religious enlightenment is its relationship with the Supreme Consciousness that we know as Buddha.

In other words, to achieve enlightenment one has to study the principles of this world and work toward the state of higher consciousness that a Buddhist would call Buddhahood. In this respect it might be said that enlightenment is an impossibly great ambition—that we can never attain full enlightenment even if we spend our whole lives on the quest. No matter how hard we strive, it is impossible to learn all there is to know about the Real World. We cannot hope to draw near to the Buddha consciousness without spending eternity in trying.

But having said this, it should be pointed out that there are degrees of enlightenment that make it possible for everyone to attain enlightenment in accordance with their spiritual stature. It is quite feasible to reach the ultimate stage of enlightenment while still occupying a body in this world.

Let us look at the different stages a person may go through on the way to achieving the ultimate stage of enlightenment. Of all the religious leaders in history, the one specifically remembered as having concerned himself the most with the question of enlightenment was Gautama Siddhartha, the Shakyamuni Buddha, who lived in India more than 2,500 years ago and who is known to most people in the West as simply Buddha, or the Buddha (and sometimes Shakyamuni). Many books describe the spiritual stages of development he went through from the moment when, meditating under the Bodhi tree (the "Tree of Enlightenment," in English often referred to as a bo tree or pipal tree, *Ficus religiosa*), he attained enlightenment and became a Buddha until the time he entered Nirvana at the age of eighty, under the Sala trees outside the castle of Kushinagara. Regrettably, however, most of these textbooks contain only fragments of his message and fail to convey to us the full revelation of his enlightenment.

The world of mind is mystical. Ten years have passed since the door opened and I became capable of communicating with my subconscious. During that time, I have been able to relive the lives of the spiritual masters of the past and know what they thought and felt, what they did, and what they experienced. I know everything that Shakyamuni experienced as he meditated under the Bodhi tree, and although it happened 2,500 years ago I can feel it as if it is happening now.

This chapter focuses on the enlightenment of Shakyamuni while also considering the relevance of enlightenment in the world today. It comprises what I intend to be a record for posterity of all I have learned about how to achieve enlightenment—it is a gift, a legacy, from the past, and a source of hope for the future.

2. Why Become Enlightened?

Why should anyone seek enlightenment? Even if one manages to achieve enlightenment, what does one hope to gain by it? To answer these questions we must consider the purpose and mission of all human beings.

To start with, why are humans born in this world? Why are we given the physical form we have here?

Before you were born as a human, you lived a life of total

freedom as a spirit in the heavenly world. In heaven, there is no need to eat to stay alive, there is no death. There is no need to work for money, and no fear of being cast out into the streets. There is no period of suffering inside a mother's womb for nine months or crying as a baby out of bewilderment and disorientation. There is none of the sexual confusion brought on by adolescence, and no conflict between parents and children. There are no financial worries, there is no need to work for others. There is no compulsion to be with people you dislike, and there is no occasion ever to suffer the anguish of parting from loved ones. There is none of the discomfort of aging or the pain and distress of illness. There is none of the misery of watching yourself grow old and being cast off by your children or grandchildren, none of the deep sorrow that comes with losing your spouse, none of the awful apprehension of sensing your own impending death. The heavenly world is a place that knows neither fear nor pain.

In heaven, people's minds are transparent, like glass, and anyone can see into them. Because there are no secrets there, there is no discord and everyone you meet is a true friend. People live in an atmosphere of love and harmony.

In the form of spirits, people are able to decide how they will look to others, and whenever they want something, they have only to wish strongly enough for it to appear there and then. If they wish to increase their understanding of the Truth, they are free to study at their own particular levels.

The spirit inhabitants of hell, in total contrast, can never be born on Earth. Their minds are filled with war and destruction, and they are for that reason forbidden from taking on human form.

So what qualifications are required in order to be born on Earth? First one has to have reached at least the Posthumous Realm (the Astral Realm) in the fourth dimension. One has to have awakened to the fact that as an inhabitant of the heavenly world, one is a spiritual being and a child of Buddha (God). This means that, in order to be reborn, one has to have reached the stage where one is repentant of any wrongs one may have committed.

From this it may be appreciated that for those in heaven it is something of a trial to be born as a human. But for those who have spent many years in hell and have finally managed to win escape through years of repentance, becoming a human being offers the chance of a new start.

This world of ours is thus something of a training-ground. For the spirits who live freely in the heavenly world, the act of becoming incarnate represents a severe test of their spirituality and Buddha-nature. The third-dimensional Earth represents a place where their spiritual awakening can be thoroughly examined to check that it is authentic. It is easy to believe in Buddha when living as a free spirit—but how many of the rules of the fourth dimension and above can be grasped when bound by the rules of life in the materialistic third dimension? Can you sense the power of Buddha? Can you perceive Buddha's power at work in the world? The spirits are tested individually and exhaustively, and it is only after they have managed to pass this scrutiny that they are able to progress to a higher dimension than the one from which they came.

The spirits who, after long repentance in hell, have at last managed to reach the lowest level of enlightenment and can recognize themselves as children of Buddha are reborn on Earth determined to prove themselves good human beings. All too often, however, they are overwhelmed by the powerful influences of the materialistic third dimension and abandon themselves to desire. Unable to return to heaven, these wretched beings plummet into an even darker hell, nevermore to recollect that they are children of Buddha.

To this extent the materialistic and ultra-physical three-dimensional world is a very harsh testing place—but at the same time it offers the possibility of salvation. It provides an opportunity to meet beings from all dimensions, beings it would be impossible to come into contact with in the Real World. You might encounter a Great Guiding Spirit of Light in human form, or meet a person whose heart is in contact with Satan in hell.

Everyone is given an equal chance. We are all born in the form of innocent children and offered the same opportunity to remake our lives. What enlightenment offers us is the chance to start our lives afresh.

3. The Path to Enlightenment
How should we go about achieving enlightenment? The word

enlightenment implies increasing the light of one's spirituality and Buddha-nature while simultaneously letting a new and brighter light shine into one's life. But in considering how it may be achieved, we may be surprised to realize that an infinite number of paths are open to us.

Those paths include the many ascetic disciplines we can follow to increase our spirituality. Buddhism is one, of course, but Christianity, Shintoism, Confucianism, Taoism, and Islam all offer their own forms of spiritual practices. Indeed, because of this multiplicity of methods there are those who, seeking the path to full awareness, to Buddha, are lost in the forest of faiths. They become confused and cannot decide which method of spiritual discipline they should adopt or which religion should be considered the true religion.

All the great religions of the world are manifestations of different aspects of the light of Buddha (God)—I mean those that have been the constant inspiration of people for hundreds or thousands of years (in this description I do not include the recently inspired faiths of various kinds). The lives and works of the founders of these great religions have been revered by their adherents for so long because they all shine with the light of Buddha. The only difference at all between them is that the light they purvey is tinted by the period, the civilization, and the way of life in which their teachings were promulgated.

This said, however, the teachings of the past are exactly that, and what is needed now is a new teaching for a new age.

The advent of a new message is badly overdue, and we must also search for a new way in which to instruct people.

Active enlightenment is a method of touching Buddha's mind with one's own. It is to search for the way, to live one's life through the mind of Buddha, constantly pursuing a pattern of life that follows the Truth.

To this end, we have been given the Eightfold Path and the notion of the stages in the development of love. If you wish to search for the path to enlightenment through Buddhism, the Eightfold Path as detailed in this book in Chapter Two offers a day-by-day guide. It preaches the unchanging Truth; anyone who follows its teachings will undoubtedly progress—but do not expect to master it all in one lifetime, for it is a path that can take all eternity.

Beginners on the Eightfold Path should concentrate on Right View and Right Speech, which alone may take anything from five to ten years to master. Once the student feels these qualities have been attained, the next qualities to devote one's life to are Right Action and Right Livelihood. When these are mastered too, the student will have achieved a level of enlightenment that puts him or her in the Light Realm in the sixth dimension.

Later on, mastering Right Thought and Right Effort, the student is on the threshold of true life in the spirit. No matter what hardship may come one's way, one will have the strength to stand firm. One's mind will become as unbending as steel and will have reached what is known in Buddhism as the state of Arakan. This is a state of mind which lies at the gateway between the Light Realm in the sixth dimension and the Bosatsu Realm of the seventh dimension. Reaching this level means that the student has achieved a considerable degree of development. But if one is still upset by trivial criticisms or if one flies into a rage over nothing, if one craves power or fame, these are all indications that one has still not reached this level.

There are a fair number of religious leaders in the world at the moment, at work both where I live in Japan and elsewhere, and in considering them we should look first at what they preach and how they practice. Some are obsessed with psychic powers; others take advantage of people's weaknesses, deceiving their followers, telling them they will go to hell or that they will be punished, and threatening them in order to appropriate their money. Such people have yet to reach the state of Arakan. The state of Arakan is the first step in becoming a Bosatsu of Light, and people whose minds are filled with a lust for power, fame, money or sexual gratification, with anger, hate, or bitterness are not true leaders sent from heaven.

To achieve enlightenment, it is a fundamental prerequisite to ensure that your mind is not lured away by worldly desires, that your heart remains pure, that you communicate with your guardian spirit, and develop the ability to see into other people's minds as if they are your own—in other words that you aim to attain the state of Arakan. Without mastering these initial stages there can be no further spiritual learning, no more enlightenment. The crucial first target must be the state of

Arakan. In the enlightenment beyond this state lies the world of Truth, real awakening, profound enlightenment.

4. Nyoshin

Nyoshin is the name of the state of enlightenment that is more profound than that of Arakan. By reaching the state of Arakan a student will have achieved an unswerving faith in Buddha, an intensely steady mind that cannot be swayed by the distracting breezes blowing in from the world. The student will have received guidance from his or her guardian spirit, and will readily see into the minds of the people he or she comes into contact with in daily life. In other words, the student will have reached quite an advanced stage of spiritual development, and as a holy person will be able to teach others.

Even in the state of Arakan, however, there is still a danger of falling from grace. For although to reach this state at all one will have reflected on one's life and be able to communicate with one's guardian spirit through the subconscious, one still will not have full knowledge of the minds of the Bosatsu of Light. One will still not have sufficient knowledge of the magnitude, profundity, and diversity of the Truth, and, therefore, there remains considerable potential for one to fall prey to misguided or false teachings.

In the inhabitants of the realms of Tengu (long-nosed spirits living in mountains that possess spiritual powers) or Sennin (hermit wizards) in Minor Heaven, the state of Arakan merely produces a primitive godly force that manifests itself in clairvoyance or spiritual phenomena. It is always important, then, to strive to deepen one's own level of love and enlightenment, and never to treat learning the Truth lightly.

There is yet another way in which people all too easily lapse from the state of Arakan. Let us liken the state of Arakan to a piece of steel that has just had its surface oxidation scraped off and is shiny for the first time. It has no coating of anti-corrosive, so unless it is polished regularly the rust will return again. That is how you need to keep your mind polished brightly. If you fail to notice that your mind has started to rust and go around thinking that you are a great teacher and have mastered enlightenment, you will be in great danger.

To continue the metaphor, when your mind is bright and shiny, its surface is very smooth and slippery and can deflect any evil influences that may try to attack it. But once it starts to rust, its surface becomes rough and pitted, allowing all kinds of things to take hold on it. Eventually, demons will come and drive a spike into the pitted rust, and from it lower a rope down into the depths of hell to allow all kinds of creatures to work their way up to you. These creatures may be lost souls, animal spirits, or even Satan himself.

There are people, then, who despite having managed to reach the state of Arakan and become religious leaders, allow evil into their minds and go on to confuse or mislead the people of this world. This is an insidiously dangerous trap, and to avoid this it is vital to keep your mind free of rust at all times. Unless your mind is shiny and polished, there is no way of knowing when it may provide a foothold for a rope to be lowered down to hell. Struggling to pull the spikes free merely results in more spikes being driven in, one after another, and once that happens all your efforts are bound to come to nothing. It will take more than a simple spiritual cleansing or exorcism to drive away the malicious spirits once they have taken a firm hold. As long as the rust remains in your mind, the demons will find a way to creep back in.

It is crucially important, therefore, to keep your mind polished. If possible, it should be coated with a hard surface layer of anti-corrosive. With rust-proofing like this you will be enabled to reach the next higher level of enlightenment.

The state of enlightenment that lies above Arakan is known as Nyoshin. In this state, one is able to receive guidance from spirits who hold a rank higher than that of one's guardian spirit—one becomes able to communicate spiritually with the Guiding Spirits who belong to the Bosatsu (angelic) Realm in the seventh dimension or higher. When you reach this state, your spirit is virtually indomitable; except under the most extreme conditions, evil can no longer assail you. Such new power derives directly from guidance received from the Nyorai and Bosatsu, which strengthens your light and makes it impossible for evil to come near.

At the level of Nyoshin your mind always remains humble and never turns to arrogance. Your main desire is to help others and be useful in the world, to search for ways to assist those who have gone astray. The main reason people lapse from the state of Arakan is that they become conceited—but once the indomitable state of Nyoshin has been achieved, one ceases to think of profit or self, and one's mind remains forever calm. It is at this stage that one finally masters Right Mindfulness and Right Concentration.

There is another side to Nyoshin. As you approach the state of Kanjizai, the upper stage of Bosatsu, you will find that you can empathize with people who are hundreds of miles away. On reading just a person's name, for example, you will in the same instant be aware of their present state of mind, their worries, what spirits possess them, and their previous and future incarnations, despite the fact they may be living on the opposite side of the globe. But you should take care not to fall into the trap of clairvoyance for your own selfish purposes. Never relax from deepening your love and polishing your mind.

5. Kanjizai: Limitless Perception

Nyoshin refers to people who have reached the level of Bosatsu and are conversant with the secrets of the spirit world. Some of them may even be able to communicate with the Nyorai Realm. Nyoshin, the stage beyond the state of Arakan, is also the state of Bosatsu. The term refers not only to the enlightened individuals of this world but also to those of the Real World.

It is a mistake to assume that the spirits in the next world know everything; that is not true. As with us, their level of knowledge and understanding varies according to their degree of perception and enlightenment. A good example of this is in their ability to foretell the future. Although there is a certain difference in degree, all the spirits of the fourth dimension and beyond have the skill of prescience and are able to foretell future events. But the accuracy of such foreknowledge is often questionable when applied to people living in the third dimension due to discrepancies in time and place.

There are two basic difficulties that account for this. First, there is the fact that future occurrences can be divided into two categories, "fixed" and "fluid" events. Fixed events are those that have been decided in the high spirit world and cannot be changed except under the most extraordinary circumstances. Fluid events, conversely, are those that appear likely to happen if things go on as they are; they comprise more of a forecast than a guaranteed prediction. As a result, fluid events can be altered through the efforts of people on Earth or at the intervention of the guardian and guiding spirits, and because of this, the prophecies of the spirits in heaven can be proved wrong.

The second difficulty is that, depending on its degree of enlightenment, each spirit is better at some things than at others—including the ability to foretell the future. Generally, the higher the level the spirit has achieved, the more accurate its powers of prescience, but some spirits actually specialize in foretelling the future and their predictions naturally tend to be more reliable.

I must now describe the state of enlightenment that lies

beyond that of the Nyoshin and which is known as the Kanjizai. The opening line of the Buddhist Heart Sutra can be translated in the words, "When the Kanjizai Bosatsu's spiritual contemplation had brought him to a very deep state, he was able to open the treasure house of his subconscious." The reference here to the Kanjizai Bosatsu is not to a particular person but to a spiritual state that is reached through long training.

A Bosatsu is the state of a soul that has passed through the Theravada stage of self-development and reached the Mahayana stage, in which its aim is to achieve salvation for the whole of humankind. Even though such souls have reached the state of Bosatsu, it does not mean that they are free of the torments and suffering of humanity or that they are always in a condition to utilize their divine powers. Progressing in their studies, however, they finally reach the highest level of enlightenment at the Bosatsu level, which is known as the state of Bonten (*Brahma* in Sanskrit). Once this has been attained, they can use their divine powers consistently despite any personal problems or illnesses they may be suffering from. The Kanjizai Bosatsu are at the same stage as Bonten and dwell somewhere between the Bosatsu and Nyorai Realms, so they can be described as inhabiting both the seventh and the eighth dimensions.

The Kanjizai Bosatsu are known as Avalokiteshvara in Sanskrit and as the Kanzeon Bosatsu in traditional Japanese Buddhism. Although they are not necessarily able to use them to the full, the Kanjizai Bosatsu are nonetheless able to employ all of the six divine powers: Tengen (Heaven's eyes), Tenni (Heaven's ears), Tashin (other's minds), Shukumyo (fate), Jinsoku (divine feet), and Rojin (hidden power).

Tengen is the power of spiritual sight. It not only gives the Kanjizai Bosatsu the power to see people's auras or the spirits that are possessing them but also allows them to see into the Real World.

Tenni is the power to hear the voices of the spirits in the other world and includes the ability to receive spiritual messages.

Tashin is the power to look into other people's minds and understand what they are thinking.

Shukumyo is the power to see into the future both of oneself and of others. It includes the power to read people's past lives in the same way.

Jinsoku is the power of astral travel. The Kanjizai Bosatsu have the ability to leave their bodies in this world while they travel in the spiritual realms; alternatively, they can teleport themselves anywhere in this world.

Rojin[1] is the state that prompted Confucius to proclaim that "Should I do whatever I wished it would not exceed the limits of morality." It is the power to surpass all desire, which means that, even though the person might have acquired spiritual power and take no thought for self, he does not cease from removing the rust from his mind or striving to achieve spiritual progress.

The Kanjizai Bosatsu possesses all these powers. It represents a step above that of the Nyoshin who "merely" have the power to read the minds of a number of people simultaneously and to participate in the experiences of someone who is a long distance away.

6. One Is Many—Many Are One

It is now time to describe the enlightenment attained when Right Thought, Right Effort, Right Mindfulness, and Right Concentration of the Noble Eightfold Path and love incarnate of the stages in the development of love have been achieved. In other words, I mean to talk about the enlightenment of the Nyorai Realm.

Up to the state of Bosatsu, the form of the soul is restricted, by and large, to the shape and form of the human body. Spirits are essentially a shapeless form of energy, intelligence without form, but after eons of reincarnation and countless lives as humans, many spirits find themselves bound to the human form and lose some of their intrinsic freedom-of-function. So, the inhabitants of the Bosatsu Realm in the seventh dimension engage in their studies within the human form. Most of the Bosatsu are unable to think of themselves in anything other than human terms—they need to think that they have two hands, two feet, wear clothes, have a certain style of hair, and have an individual face. They cannot relax unless they take human form. Virtuous in the extreme though they may be, and with highly developed powers of leadership, their power is nonetheless limited by this need for human form.

The situation is rather different in the Nyorai Realm in the eighth dimension. The inhabitants of the Nyorai Realm know that they are not human-shaped spirits. They retain their memories of having lived as humans, of course, but they are aware that a spirit is a form of conceptual energy, a shapeless "package" of light. Not only do they understand this at an intellectual level but they acknowledge it in the way they live.

If a human psychic were to leave his physical body and travel to the eighth dimension, what would he see? To make things easier for him to comprehend, the Nyorai would appear to him in the human form they had used when they were themselves living on Earth. They would invite him into their homes where they would offer him coffee or wine to drink, but the coffee or wine would have a taste and aroma that people only dream of on Earth. When the human returned to this world, he could give his friends a description of his impressions. "The Nyorai Realm of the eighth dimension is a truly amazing place," he might well say. "The inhabitants are godlike in appearance, the streets are paved with rubies, and the buildings are scattered with diamonds. The drinks have an aromatic fragrance, the tables are made of shining marble, and the pillars at the corners of the rooms are made of pure crystal."

It was an impression like this that the famous eighteenth-century European theologian and psychic Emanuel Swedenborg (1688–1772) presumably meant to report, although his description shows that he still had some way to go in attaining true spiritual awareness. If he had looked more carefully at his surroundings, the jewel-encrusted buildings and streets would have disappeared, leaving only the Nyorai smiling in front of him. If he had looked even closer still, the Nyorai themselves would have disappeared and he would have been faced with a giant ball of light. The rubies, diamonds, and other jewels were produced to delight the visitor from Earth, as an attempt to translate the world into something that would be understandable in three-dimensional terms.

The enlightenment of the Nyorai Realm permits its inhabitants to recognize themselves as formless entities, and those who have reached this stage while still living on Earth will be aware of the law that pertains in the Nyorai Realm.

The law of the Nyorai Realm states that "One is many—many are one." In the world of Nyorai there is no objective way of recognizing "one," so it might just as easily be "ten" or "ten thousand." In the same way, "one thousand" might just as easily be "one." These numbers are not an objective fact, they are merely numbers in the mind; only the unifying consciousness knows the true state.

Simplifying matters a little, let us say that if a being from the Nyorai Realm had ten jobs to do, it could become ten people. If it had ten thousand jobs, it could become ten thousand people—but even if it did become ten thousand, it would still only be a single consciousness.

A modern Japanese philosopher of the Kyoto School, Kitaro Nishida (1870–1945), became aware of this law of the Nyorai Realm after many years of study. Of course, he was himself an inhabitant of the Nyorai Realm and his subconscious provided him with the knowledge of that world.

The Nyorai Realm of the eighth dimension is a place where the utterly contradictory unification of self is possible. It is a world in which things that are visually different and contradictory may be intuitively unified as one. Kitaro Nishida seems to have gained partial comprehension of this realm—partial enlightenment—during the course of his life.

7. The Enlightenment of the Realm of the Sun

The enlightenment of the Nyorai Realm is an understanding of the principle "One is many—many are one," a concept generally beyond the reach of human comprehension. It is awareness that spirits are particles of Buddha's light, shapeless bodies of energy, formless intelligence. The Nyorai also marks the practical limit of enlightenment achievable by people on Earth. This is illustrated by the fact that out of a total population of souls in the terrestrial spirit group numbering tens of billions, fewer than five hundred live in the Nyorai Realm. It also proves how difficult it is to achieve this level of enlightenment.

The enlightenment of the Nyorai goes beyond the simple elements of good and evil, and reaches the stage of integration and sublimation. To achieve such enlightenment it is not sufficient merely to improve oneself through human discipline and effort; also required are ultra-lucid reasoning powers and the most penetrating wisdom in order to be able to understand the vastness of the events that take place in the universe and the laws that govern them.

What kind of people can achieve this rare state? Let us look at some people who have managed it in the past. There is Shotoku Taishi, or Prince Shotoku (574–622), regarded as the father of Buddhism in Japan and the founder of the state itself. There are also Kukai (774–835), the priest who founded the Shingon sect of Buddhism, the philosopher Kitaro Nishida (1870–1945), and a few others.

Although the Nyorai Realm is inhabited by fewer than five hundred souls, it is divided into four sections. The lowest is the Realm of Bonten (Brahma), which marks the border with the seventh dimension and which is inhabited by approximately forty Nyorai. Next comes the Semi-Divine Realm containing approximately one hundred and twenty souls. Then there is the Divine Light Realm and its two hundred and eighty souls. The highest section is known as the Realm of the Sun, although in a broader sense the Sun Realm includes the whole of the Cosmic Realm of the ninth dimension. In its narrower sense, however, the Realm of the Sun lies on the border between the eighth and ninth dimensions and is inhabited by around twenty Grand Nyorai.

The Grand Nyorai who live in this Realm of the Sun include three Shinto gods—Ame-no-Minakanushi-no-Kami,[2] Ame-no-Tokotachi-no-Kami, and Kamumusuhi-no-Kami— the Western Christian saints Augustine and Thomas Aquinas, the Taoist leaders Lao-tzu and Chuang-tzu, the apostle of

Moism, Mo-tzu (sometimes called Mo Ti), the ancient Greeks Apollo, Socrates and Plato, and the Buddhist leaders Ashuku Nyorai and Yakushi Nyorai. In the Divine Light Realm closest to the Realm of the Sun lives Islam's Muhammad.

What is it that these inhabitants of the Realm of the Sun share in common? Simply, they have all reached a degree of enlightenment that is impossible for a regular human to achieve. It is not a state that can be reached through human endeavor, and as a result they have become divine. The Realm of the Sun marks the beginning of the realms that go beyond human enlightenment, the realms in which the inhabitants are all Grand Spirits. Any one of them has the stature to become the founding focus of a major religion on Earth.

One of the duties they have as beings who have gone beyond the realm of human enlightenment is to be personally involved in planning the evolution of humanity. They help the Grand Spirits of the ninth dimension and are responsible for designing and creating new civilizations, new religious revolutions, and new eras in human history.

8. Shakyamuni's Enlightenment (1)—The Great Awakening

Before I go on to describe the enlightenment of the ninth dimension, I must reveal how the enlightenment of Gautama Siddhartha, the Shakyamuni Buddha, occurred in India more than 2,500 years ago.

At the age of twenty-nine Gautama Siddhartha left the home of his wealthy parents and set out in search of enlightenment. For six years he led the life of an ascetic, but abandoning these austerities he finally attained enlightenment under the Bodhi tree at the age of thirty-five. After a week's continual meditation, one morning, from around one o'clock, he could finally enter a deep trance, in which he experienced a great awakening. He might well have described the enlightenment that came to him then in words much like those below.

For many years, I sought spiritual enlightenment through the mortification of the flesh. I denied myself food and water and submitted my body to extremes of pain and deprivation.

Six years have passed since I left my wife Yashodhara and my son Rahula and, ignoring the pleas of my father to succeed him, fled from Kapilavastu. When I lived in my father's palace, I had a powerful physique and was an adept at both the military and literary arts—but look at me now. My ribs protrude, my eyes are sunken, and I look like a skeleton. If such mortification of the flesh is a necessary part of spiritual training, why are we born into human flesh at all? If it is the will of Eternal Buddha[3] for us to maltreat our bodies in this way, those who commit suicide must surely achieve the highest enlightenment.

But what is the result of suicide? Because the law of the universe corresponds to the chain of cause and effect, if we sow evil, evil is what we shall reap. If our suicide causes fresh anguish, unimaginable suffering will await us in hell. Torturing our bodies in this way is simply a form of slow suicide. The state of Buddhahood is one of tranquility, but there is no tranquility in the austerities of asceticism, nothing that will lead to enlightenment. All that I achieved through six years of asceticism was a frightening appearance and penetrating eyes. All that my discipline brought me was a grim face and eyes that can pierce like arrows—but nowhere is there any trace of love, of compassion. If there is no tranquility, no joy in our own hearts, how can we show true kindness or true compassion to others?

But what exactly is meant by "joy," by "happiness"? When I was living as a prince in Kapilavastu, everyone did my bidding—I had all the money, all the women, all the material goods that I could ask for. But was I happy? No, all I felt was a lukewarm languor. My heart was always starving; my heart was always thirsty. Entrapped in the mesh that is the network of other people's desires and expectations, my heart was in a permanent state of resentment and confusion. I knew that in due course I would become king and would doubtless have to lead my people to war with a neighboring country, and that the inevitable result would be bloodshed and death.

If we search for earthly power and fame, all we

find is emptiness. My days in Kapilavastu were any-thing but happy; I was spiritually unsatisfied, and felt nothing but restlessness and impatience. Human hap-piness is not to be found in inactivity and inertia but in daily spiritual progress. For the children of Buddha, true joy lies not in worldly success but in the better-ment of our souls and Buddha-nature in accordance with the will of Eternal Buddha.

Real enlightenment for the children of Buddha does not lie in the elegant comfort of palace life any more than it can be found in the extremes of asceticism. Real enlightenment, true happiness, genuine peace can no more be found in the extremes of physical pleasure than they may in the extremes of physical pain.

The true way to enlightenment avoids extremes and corresponds to the middle path. A balanced life takes the middle path, and brings with it a world of perfect harmony. The life that all people seek is a life of harmony, and to achieve it they must abandon extremes of pain and pleasure and stick to the middle path. Follow the precepts of the Eightfold Path—Right View, Right Thought, Right Speech, Right Action, Right Livelihood, Right Effort, Right Mindfulness, and Right Concentration—and you will find the true king-dom of the mind, the true kingdom of Buddha.

True human happiness lies in a measure of spiritu-al joy and progress every day, and that happiness is

increased to the heights as we master the profundities
of the Eightfold Path.

9. Shakyamuni's Enlightenment (2)—Passing into Nirvana

I have been outlining Shakyamuni's state of mind when he achieved his first awakening at the age of thirty-five. Even as I have been writing, I have been seeing what he experienced 2,500 years ago when he reached enlightenment. If my pen was left to its own devices at this point, it would probably write enough to fill a whole book. But I must now hurry forward forty-five years to the day that Shakyamuni entered Nirvana at the age of eighty. I must again describe what he felt and what he learned as it might have been expressed by his own voice.

He lay on his right side under the Sala trees in Kushinagara, his right arm folded under his head as a pillow, and his left hand held to his aching stomach. As he entered Nirvana, his thoughts turned to his life and disciples.

In the forty-five years since I achieved enlightenment I
have sought goodness and preached the Truth, but
now the time has come for me to bid farewell to my
physical body. All things are transitory, and I no
longer feel any attachment to you, my weary old body.
For more than forty years I have been able to teach
people the Truth of Buddha and show them the way to
live an ideal life, and these teachings have grown to
become my true body.

I must also thank you, my disciples, for staying with me over the years, looking after my personal needs and helping to spread the word. Thanks to your efforts the community of my followers, the Sangha, now numbers more than five thousand and there are hundreds of thousands of lay followers throughout India You have battled against religious persecution and eluded our enemies to spread the Truth of Buddha. Without you, none of this could have been achieved, and I pray that you will continue with the task after I am gone.

I look forward to meeting you, Shariputra. You left the physical world several years ago, but I am looking forward to meeting you there again and enjoying another of our talks. You were a great help in my task, and it is only natural that you should have become known as the "first in wisdom." You were always a good listener, and that made it very easy to preach to other disciples, as well. Sometimes you made me laugh with your stupid questions, but by asking them you helped those in the audience who were too shy to ask for themselves.

And Mahamaudgalyayana, you were called the "first in psychic powers," and even I was unable to hold back the tears I felt when I heard that you had been killed by unbelievers. I can see you even now, coming to meet me, riding on a shining cloud.

Mahakatyayana, you were the "first in logical

argument" and were always able to explain my teachings in a way that could be understood by all. I know
you will continue to spread my teachings in remote
areas after I have gone. You should go to the Avanti
region in western India and do your work there.

Subhuti, you were the "first in understanding the
concept of emptiness." You never allowed yourself to
become enamored of material things and very quickly
managed to grasp my teachings of selflessness and
emptiness. Go on with your studies: never slacken,
never stop.

Aniruddha, I once lambasted you for falling asleep
while I was preaching, and as a result you went on
meditating without pause, night and day, until finally
you lost your eyesight. Fortunately, you learned to see
through your spiritual eye and became known as the
"first in clairvoyance." But you, once so young and
clean-cut—your hair is turning gray now.

Purnamaitrayaniputra of the Shakya clan, you
were always a clever man and soon became recognized as the "first in preaching a sermon." You and
Purna, who is also planning to travel to the west to
preach, will be well-matched rivals.

Mahakasyapa, you will not return here for another week and will, therefore, not be able to witness my
passing on. You will be furious with Ananda who inad-
vertently fed me poisonous mushrooms and so has-

tened my departure from this world. You will try to have him expelled from the order and will weep bitter tears. You are known as the "first in discipline," and were always becoming involved in the minutiae of our religious practices. After my death, though, you should do away with the minor rules.

Upali, you were always the "first in keeping the rules of the order." You have always done your work well and been courteous to all you have met. You were born of a lowly caste, but now you mix with nobles without in the least feeling daunted. You have made tremendous progress and I am proud of you.

Rahula, my true son, you strove in secret under Shariputra and as a result became called the "first in unnoticed spiritual practices." It was thought that you would succeed me, only you died so young. I could not do anything for you as your earthly father, but I hope that you are living in joy in heaven.

Jivaka, you are the greatest doctor in the world. You have cured my sicknesses time and again in the past, but this time I am beyond even your skills. All things are transitory, and just as you cannot stop the flow of the river, my life on Earth cannot be extended any further.

Finally, my dear Manjushri, I have a message for you. At responding with wise words you are as sharp as a sword. On that occasion when even my ten fore-

most disciples were unable to answer the challenges of the heretics and the misguided laymen, you were able to defend our order. You were able to hide your spiritual identity of Ryushu Nyorai while living the life of a young Bosatsu. When I have gone, you should take an elephant and ride to your homeland in southern India in the guise of a man. There you must establish disciples of your own and spread my teachings. You will be remembered in history as the founder of Mahayana Buddhism. In the future, when I am reborn in an Eastern country, you shall be with me as my wife.

Oh, when I think of you, my beloved disciples, I worry about what will happen to you after I am gone. But I want you to remember that although I will not be with you in person, my teachings will remain for thousands of years as nourishment for human souls.

Disciples for all time, I want you to remember these, my last words. My life is like a full moon, and when I am no longer with you it will not be that I no longer exist but just that the clouds of death have come between us. I will still be there on the other side of the clouds, shining as brightly as the moon. Life shines forever and knows no end.

When I am gone, I want you to nourish yourselves with the teachings I have preached for the last forty-five years. You no longer need anyone to light the way; you have a light in your souls that will shine out ahead of you. Grow strong on the words I have taught you while using them to save others.

After I am gone, remember all I have said about the inner light, and live your life in the Law. These are my last words. As the various events of this world take place and go by, do not rest until you attain your enlightenment.

Although Shakyamuni was hardly able to voice the words as his death approached, his disciples had opened the doors of their minds and were able to hear his innermost thoughts, which they thereafter recorded in the Nirvana Sutra.

10. The Enlightenment of the Ninth Dimension

Shakyamuni's enlightenment surpassed even that of Jesus Christ and was the highest ever attained by a human being. Unfortunately, however, he was unable to pass on what he had learned about the Grand Cosmos in its entirety during the forty-five years he was alive on the Earth. Hardly any of his disciples were able to reach the Nyorai level of enlightenment while they were alive, and it was thus very difficult for them to attempt to understand the creation of the world or its multidimensional structure.

In those days, India was comprised of many warring states, and even if Shakyamuni had preached the truths that surpassed the knowledge of that time, it would not have helped to save people's souls. For this reason, he concentrated on teaching the Noble Eightfold Path in an effort to raise the consciousness of the people to the state of Arakan.

But to achieve the enlightenment of the ninth dimension, people should essentially fulfill three conditions.

They should first have a complete understanding of as many areas of life as possible, and of the Laws; this will allow them to talk spontaneously to any kind of audience.

Second, they should have full knowledge of how the human species came into existence—a concept that includes knowledge of the composition of the universe and the history of the world.

Third, they should be fully aware of the multi-dimensional universe in the fourth dimension and above.

Shakyamuni was himself a master of the kind of spontaneous preaching indicated in the first condition above. His knowledge of creation was obtained when he gained enlightenment under the Bodhi tree and his spiritual body became one with the Grand Cosmos. His knowledge of the multi-dimensional universe, the Real World, is expressed in his Law of Karma.

The enlightenment of the ninth dimension naturally includes total mastery of the six divine powers, involving the ability at once to perceive humanity's past, present and future. Shakyamuni realized the danger of people's worshipping supernatural power for its own sake, however, and for that reason made very little use of his powers other than the ability to read minds.

In Chapter One, I stated that there are ten Great Guiding Spirits of Light who embody the enlightenment of the ninth dimension. It is appropriate here to list them and their present roles and responsibilities (as of 1994):

Major Heaven

1. **Shakyamuni** (El Cantare): The most powerful spirit of the terrestrial spirit group. *The creation of a new age and the building of a new civilization.*
2. **Jesus Christ** (Amor): *Deciding the guiding policies of the heavenly world.*
3. **Confucius** (Therabim): *Planning the evolution of the terrestrial spirit group and interstellar relations.*
4. **Manu**: *Racial problems.*
5. **Maitrayer**: *The refraction of Buddha's light.*
6. **Newton**: *Science and technology.*
7. **Zeus** (the original person behind the myth): *Music, art, and literature.*
8. **Zoroaster**: *Moral perfection.*

Minor Heaven

9. **Moses** (Moria): *Chief commander for the dissolution of hell and controller of miraculous phenomena.*
10. **Enlil**: *Guidance of the Realm of Sorcery (Arabia), the Yoga Realm (India), the Sennin Realm (China), and the Sennin/Tengu Realm (Japan). (Gods of vengeance)*

In this chapter, I have described the various levels of enlightenment up to the ninth dimension, and although there is such a thing as the enlightenment of the tenth dimension, the inhabitants of that dimension are the consciousnesses of the Grand Sun, the Moon, and the Earth respectively, and it is not a state that can be attained by a mortal spirit. To try to give more detail to it, however, it could be said to correspond to a state of enlightenment from which all human elements have been eradicated: a huge ball of light with a powerful sense of purpose.

Notes

1. The term Rojin refers to an ability to free oneself of all worldly delusions. It is achieved through daily contemplation and self-examination, and should accordingly be considered a state of wisdom rather than a psychic power. Once this state has been reached, it is nonetheless possible for a person to live an outwardly normal life while being the possessor of extraordinary psychic powers.

2. Ame-no-minakanushi-no-kami (literally, God Ruling the Center of Heaven): in the *Kojiki* ("Records of Ancient Events," the first history of Japan, compiled in 712) this god is said to be the supreme power that controls the land of the gods. After an extensive search of the spirit world, however, I have discovered that although he is head of the Japanese Shinto gods, he was a godly spirit who came to live on Earth three thousand years ago. He lived in Kyushu, in the south of Japan, and is the ancestor of the present Imperial family.

Stories telling of the creation of the world, like stories of the gods in the heavens described in ancient histories, cannot be proved to have any factual basis and it is best to consider them as being founded on the words and deeds of powerful people who lived in Japan in prehistoric times. Amaterasu-O-Mikami (the Sun Goddess) was not really a divine embodiment of the sun but was in fact a woman of high birth who was worshipped as a female Guiding Spirit after she returned to the Real World.

3. The term Eternal Buddha generally refers to El Cantare. Here it is used to denote the core consciousness of the spiritual body group of Shakyamuni Buddha.

Five: The Golden Age

1. Precursors of a New Race

The twenty-first century is now upon us—but what kind of people will emerge in the new millennium? What kind of era will it be? Many people regard the twenty-first century with a mixture of fear and expectation.

Yet portents of the new age—precursors of a new humankind—are already visible within our society. We are presently living in a transitional period, a period that will see the disappearance of much that is old and the sudden emergence of much that is new. The seeds of the new era are already with us, and it is the task of those who know what is coming to convey the knowledge to humanity.

Humankind witnessed the destruction of an ancient civilization when Atlantis sank under the ocean approximately 10,000 years ago. The end of one age means the beginning of another, and on this occasion a new civilization sprang up in Egypt, initiating a chapter in the history of humankind that has now lasted for nigh on 10,000 years. But this chapter is due to finish with the end of the twentieth century.

During this period, many cultures have arisen on various continents and in regions around the world—for example in Egypt, Persia, Judea, China, Europe, America, and Japan— and the one thing they all have held in common is that they have revered the power of the mind, they have worshipped at the altar of "intellect"; all of them have tried to understand the world around them through sheer intellect. Ours can, therefore, be called the Age of Intellect.

In contrast, Atlantis produced a culture based on reasoning, and Nyorai Maitrayer and Koot Hoomi (better known to us as Archimedes and Isaac Newton) of the ninth dimension were greatly instrumental in its formation.

The civilization of Atlantis was itself preceded by the civilization of Mu, which flourished on the continent of Mu, in the Pacific Ocean, some 15,000 years or so ago. This culture can be summed up as one based on the energy of light: It studied light energy from both the scientific and religious viewpoints, and the chief ambition of each of its citizens was to expand on his or her individual light power.

Before Mu, however, a continent called Lamudia, a landmass located in the Indian Ocean, existed more than 27,000 years ago. The people of this civilization concentrated on the development of the five senses. Whereas El Cantare (Shakyamuni Buddha) was closely involved in producing the civilization of Mu, Lamudia was mostly the work of Manu and Zeus, who between them produced a civilization based on sensation and sensitivity. The people were trained to develop their senses so that in the fullness of time the most adept were capable of distinguishing 3,000 colors and 2,500 different odors.

Even farther back in prehistory, on the continent of Myutram, a civilization prospered but then disappeared around 153,000 years ago. In those days the orientation of the Earth's rotational axis was very different from its present alignment; the landmass that we know today as Antarctica lay in the temperate zone and was known as Myutram. This continent—unlike Mu and Atlantis—was not later to disappear beneath the sea but instead, as the Earth's axis shifted, to become covered with ice so that most of the indigenous life forms perished. Some vague racial memories of this event remain within the human psyche, coloring what is thought of as the "Ice Age," and although the continent has changed somewhat in shape over the millennia, many traces of its former civilization remain trapped under the perennial ice.

The Myutram civilization was itself preceded by the Garna civilization, which existed approximately 735,000 years ago, when the African and South American continents were still connected to each other. This giant continent was known as Garna, and the culture that flourished there was founded on the principles of psychic power. A sudden violent agitation in the Earth's crust split the continent, however, and its civilization was destroyed by an associated earthquake of such terrifying power that it would register ten on the Richter scale today.

2. The Garna Civilization

It may be quite unfamiliar, but what I am describing here is not something out of a cheap science fiction magazine—it is the actual history of the world, which I know to be absolute fact

from my own proven spiritual connections. A clear understanding of what has happened in the past is essential in order
that we may gain some insight into what will happen in the
future, and the cultures that will inhabit it.

The 400 million years of human history has seen a good
many civilizations rise, flourish, and then disappear, like bubbles in a stream. There is no need for us to study of all these
cultures in detail. We should learn just enough to give us a better understanding of present and future civilizations. For this
reason I have decided to refer to the Akashic records—the history of the world that is kept in the Real World—to give a brief
account of the history of the major cultures during the last one
million years, starting with the civilization of Garna.

The continent of Garna came into being when a tremendous volcanic eruption took place on the sea-bed about 962,000
years ago, causing the bed to rise and form a landmass in the
area that now lies between Africa and South America. It ended
when a violent movement of the Earth's crust some 735,000
years or so ago finally split the supercontinent in two in a disaster of previously unheard-of proportions.

During that lengthy period, no fewer than four civilizations emerged on the continent, but it is the final one that is of
particular concern to us, and that we will call the Garna civilization. This civilization flourished for approximately 25,000
years until the continent was destroyed. Its inhabitants—who,
as we have seen, diligently pursued the knowledge and use of
psychic powers—were unlike people today. For, in those days,

the average height of men was two meters ten centimeters (nearly six feet eleven inches) and that of women was one meter eighty centimeters (nearly five feet eleven inches). Moreover, the men of that time still had their third eye; a round, emerald-green eye in the center of the forehead about two centimeters (almost an inch) above the eyebrows. This eye generally remained closed, opening only when the men were exercising their psychic powers. Without such a third eye, the women lived in fear of the power it gave the men, gradually becoming no more than vassals of their menfolk.

At the same time, women were accorded some reverence on account they had a womb. One of the Garnan folk tales told how "God made man and woman equal, as is obvious from the fact that He gave the men their third eye to enable them to protect themselves and bring safety to their people, while to the women He gave the womb to let the people prosper." In those days, a woman's womb was thought to be vested with psychic powers that enabled women to converse with the spirit world and call down a spirit to be their child. The mother-to-be and the spirit would discuss things fully beforehand to be sure that they were both happy with the arrangement, and in consequence there was no need for such a thing as abortion.

The Garna civilization was made up of eight different races, who were forever warring between themselves for supremacy. So there was a constant need for protective measures to guard against attack. The third eye could also be used as a weapon, and its color varied from race to race; eyes could be yellow, green, mauve, black, gray, or brown. Each race was further differentiated by its specific level of psychic power. The major faculty of the third eye was a power to affect the physical world, a force that we now call psychokinesis. Depending on the race, a man might be especially gifted at (for example) foretelling the future, which would enable him to give warning of any impending attack and so afford him the chance to save his people.

Unfortunately, in their quest for the Truth, the Garna people were not interested in the mind itself, preferring to concentrate instead on the development of their psychic powers. As a result, when the continent disappeared and they all returned to the Real World, most of them went to enlarge the Sennin, Tengu, and Sorcery Realms in Minor Heaven.

There have been no more people born with a third eye since the destruction of the Garna civilization—but the "frontal chakra" spoken of in yoga harks back to its existence.

The Garna civilization, a culture based on psychic power, came to a violent end, destroyed in a disaster of cataclysmic proportions when the continent was split in two. It happened one autumn evening approximately 735,000 years ago. There was a terrible rumbling roar from within the Earth, and a vast crack appeared running north to south through the center of Ecarna, the cultural capital of the country. The fault expanded rapidly until it was a hundred kilometers (more than sixty miles) long and started to fill with seawater, ominously presaging the imminent break-up of the continent. Total disaster struck three days later when a vertical shock wave ripped through the city and tore it apart. Ecarna's entire population of 300,000 perished that day.

The fault line continued to extend until it stretched for several thousand kilometers, splitting the continent in two. The process of division took tens of thousands of years to complete, but eventually the two halves drew further apart and became what we now know as Africa and South America.

In the southeast of Garna there was a city named Emilna, in which the inhabitants were particularly skilled at foretelling the future. Because of this faculty, some Emilnans managed to foresee the coming catastrophe and escaped by ship to a new, uninhabited continent that had been discovered to the south. The story of their journey is what has come down to us as the tale of Noah's Ark. But they had lost all the tools of their civilization in addition to many of the most talented people. They inevitably degenerated into a simple agricultural people, and as they did so the power of their third eye waned.

3. The Myutram Civilization

Many civilizations rose and fell on this new continent, but by far the most important was the Myutram culture, which flourished between 300,000 and 153,000 years ago. The continent was finally named after this culture. Modern Antarctica comprises some eighty percent of the original continent of Myutram, but the Earth's rotational axis was different in those days from its present orientation, and Myutram had a temperate climate. The inhabitants formed an agricultural society and cultivated a grain similar to modern strains of wheat.

In fact, the Myutram culture was unique in its study of food. Considerable research was carried out to discover what foods, and what combinations of foods, were the most useful, both nutritionally and in respect of the effects on people's physical, mental, and spiritual lives. The agronomists discovered what foods gave people a gentle disposition, what kind of meat improved people's reflexes, what dairy products should be eaten how many times a day to extend the human life span, and what types of alcohol had a positive effect on the human brain. They studied just about every possible aspect of food.

Moreover, each branch of food science had its own specialists. There were, for instance, specialists in foods that enhanced longevity, foods that increased stamina, and foods that improved the memory. All the children studied hard to become a specialist in one of these fields. The result was a huge volume of research on the relationships between eating habits and human temperament. Unlike the Garnans, however, who were warlike but took spiritual powers very seriously, the people of Myutram were peace-lovers and tended to look down on anything connected with the spirit world. To this extent, it would certainly be fair to describe their culture as the first one that stressed materialism. Although their discoveries in the links between food and human temperament were undoubtedly important, by concentrating on food in this way they neglected humanity's original mission, which is the study of the soul and self-betterment through spiritual discipline.

Many people who are today closely associated with health or slimming foods no doubt lived several lives in the Myutram civilization in previous incarnations, and have not forgotten their fascination for analyzing and experimenting with food.

The Myutram civilization reached its peak some 160,000 years ago, a time at which Moria came down to Earth in the form of a holy man named Emula. He set to work to foment a spiritual renaissance under the slogan "From a life of food to a life of the Mind." However, he was persecuted for belittling Myutram's nutritional traditions and in the end his efforts all came to nothing. Yet he did succeed in making the people sit up and think that perhaps there was more to this world than just food, and in this respect he may be called the forerunner of the modern religious movement against materialist values.

But 153,000 years ago there was a radical shift in the axis of the planet, and the climate of Myutram turned Arctic. This sudden alteration in environmental conditions was the direct cause of the destruction of the Myutram civilization.

One evening, the people noticed that the sky had turned abnormally red, quite unlike an ordinary sunset. It was as if the whole sky had been dyed with blood. They asked their scientists what was happening, but none of them was able to explain it. At about ten o'clock that night, some of the people noticed that the stars seemed to be falling from the sky, and to their horror also realized that this was not just a comet shower but that the stars themselves were changing position. The stars were certainly changing position but only in relation to the Earth's daily rotation, for in reality it was the Earth that was moving in an untoward manner. The planet spun like a ball that shoots up out of the sea, and when it came to rest its rotational axis had shifted. The effects of this phenomenon were felt quite clearly over the following months. Snow began to fall in the previously temperate Myutram, and the ground froze. This sounded the death knell for the agriculture-based civilization, and people began to starve.

Some of the people tried to hold out in a city built underground, but this, too, lasted no longer than two or three years. Unfortunately for all involved, Myutram had just entered the monsoon season when the poles moved, and the rains fell in the form of snow—more than five meters (sixteen feet) of snow fell in less than two weeks. The capital city, La Myute, was simply snuffed out by it, although some of its inhabitants managed to escape on ships. In this way, some part of the Myutram civilization managed to find its way to and settle on a new continent.

4. The Civilization of Lamudia

At that time, there was still no continent in the Indian Ocean, just an island about twice the size of Japan. It was here that several thousand refugees from Myutram settled and began to raise their families. Around 86,000 years ago, however, this land began to rise up out of the ocean and, over as short a period as about one year, it grew until the continent of Lamudia was formed in its entirety. It became the largest continent on Earth, roughly diamond-shaped, stretching 3,500 kilometers (2,175 miles) from east to west and 4,700 kilometers (2,920 miles) from north to south. Trees and plants gradually covered its surface, and the soil proved very fertile.

About 44,000 years ago, the man who was destined one day to be reincarnated in ancient Greece where he would be known as Zeus, was born in Lamudia and was called Elemaria. A man of exceptional holiness, Elemaria was a genius in writing, drawing, painting, and music, using his artistic brilliance as a medium through which to demonstrate to people the joy of living and the glory of God. As a result of his work, the Lamudian civilization became renowned for its outstanding creativity in things artistic: music, painting, literature, poetry, architecture, and sculpture. Many of the people who excel in the arts today learned their skills in Lamudia in a previous life.

After the passing of the great and holy Elemaria, Manu also came down to Earth in Lamudia, and brightly illuminated the world with his presence. Born 29,000 years ago, he was then known by the name of Margarit which means "the competitor," a name he was given for two reasons: one, he was held to rival the great and holy Elemaria, contemporarily worshipped as Almighty God; two, he taught the tribes to compete with each other through art.

Manu, or Master Margarit as he was called, infused the arts with the principle of competition. First he divided the five arts—music, painting, literature, architecture, and folk handi-crafts—between the five tribes and encouraged them all to strive for perfection. Then he organized a competition, held every three years, to pick out the finest work created during the period, and gave the winner's tribe the authority to rule the country until the next competition. The competition was restricted to the limited field of the arts, but it was completely fair and, in as much as the winner was given power to rule for a defined period, it could be said to represent the prototype for democracy. Furthermore, by teaching that art led ultimately to God, Master Margarit worked toward the unification of the Church and the state.

The Lamudian civilization disappeared suddenly approximately 27,000 years ago. It was an extremely hot summer afternoon and the inhabitants were all euphorically listening to beautiful music. As connoisseurs of the arts, people were accustomed to devoting themselves to music for two hours every afternoon, and it was during just such a break that the catastrophe occurred with terrible suddenness. The chandeliers hanging from the ceiling began to sway wildly and all the glass in the windows shattered. The huge contemporary concert hall was razed to the ground in a matter of minutes as the whole of the eastern side of the continent began to sink into the ocean.

By four in the afternoon, half of the continent had disappeared. The following morning, the sun rose at seven to shine down on an unbroken expanse of empty ocean. No trace of the continent remained—just an innumerable quantity of corpses being tossed by the waves of the Indian Ocean. The destruction of the Lamudian civilization was total; the entire population of two and a half million vanished under the sea. Good or evil, there were no survivors. But their culture was saved. The Lamudians had a colony on a continent named Moa, later to be better known as Mu.

5. The Mu Civilization

The continent of Mu in the Pacific Ocean was geologically older than that of Lamudia, having emerged from the seas some 370,000 years ago. Its shape altered several times over the millennia, but by the end of the Lamudian civilization it corresponded to a continent that was approximately twice the size of modern Australia and was located where Indonesia now stands.

People had lived on the continent for hundreds of thousands of years but in cultural terms they were not very advanced. In the north, the inhabitants subsisted mainly on fishing; those in the south lived by hunting; and those in the interior were mostly farmers. As the civilization on Lamudia flourished, its people turned their eyes toward Mu and decided to colonize it. They constructed a huge navy of sailing ships and, from around 28,000 years ago, they began to take over the major cities. Some of the natives were transported back to Lamudia to work as slaves and do the menial tasks, leaving the colonists free to devote themselves to the arts. Such a disregard for their fellow humans created dark clouds of disruptive negative energy that blanketed the civilization toward the end, and it was as a direct result of this that the continent in due course sank under the waves.

The cities of Mu may have been mere colonies, but they were thoroughly permeated with Lamudian culture—society was Lamudian in tone and form. After Lamudia was destroyed, however, a culture that was unique to Mu gradually began to emerge.

About 20,000 years ago, Escallent—who was later to be reincarnated as Zoroaster—appeared on Mu. The great and holy Escallent's teachings focused on the scientific energy of the light of the sun. His message was that the power of light could be understood in two ways: first it was a holy thing that represented the glory of God, and second it was something that was of practical use to all people.

As a gesture of reverence for light, light being holy, Escallent taught the people to put their hands together, bow, and go down on one knee whenever they saw a source of light, whether it was the sun, the moon or even the artificial light in a room. This sign of respect was passed on to Eastern culture and is the origin of the tradition of bowing in that part of the world.

To emphasize that light was of the ultimate usefulness Escallent borrowed the help of Koot Hoomi (later to be reincarnated as Archimedes and Newton) and the scientific genius of Enlil to create ways of amplifying the power of light. It was at this time that people used gigantic light amplifiers in place of generators to light their rooms, to drive their ships, and to power industry. At the center of every city was a huge triangular pyramid each side of which measured thirty meters (just over ninety-eight feet) and which shone with a silver light. The pyramid absorbed and amplified the light of the sun, then transmitted it to smaller pyramids, the sides of which measured ten meters (nearly thirty-three feet), which were located throughout the city and which in turn transferred the energy to individual pyramids, with sides measuring one meter (three feet 3.4 inches), mounted on the roofs of individual buildings.

This form of pyramid power was passed on to the civilization of Atlantis and is similar in principle to what we talk of today as pyramid power. The preparations were thus laid for the coming of the scientific age.

6. The Age of La Mu

The Mu civilization reached its peak during the La Mu era approximately 17,000 years ago, when belief in the holy nature of the sun and the use of solar technology was universal. La Mu was an earlier incarnation of Shakyamuni who was born on the Moa continent at this time. His name means "Emperor of the Light of Mu." Moa now came to be known as Mu, and the culture that had grown there was known as the Mu civilization.

Very pleased with the scientific culture that had blossomed on this continent, La Mu felt that there was now an ideal opportunity to found a kingdom of God on Earth. La Mu possessed immense psychic powers and could communicate freely with the spirits in heaven, where his main guiding spirit was Amor, later to be known as Jesus Christ. His teachings made three major points.

First was that all the people of Mu should understand that God was an entity comparable with the sun, that like the sun He poured down His light on the people living on Earth. Second, La Mu stated that all the people of Mu should live in a spirit of love and compassion, as if enfolded within the light of the sun—that their love and compassion should be visible in the way the hearts of others were filled with light. Third, he said that all people should strive to improve themselves—and by this he was not referring solely to the arts or to academic learning or to the martial arts, but above all to spiritual perfection, which was the goal of his teaching.

It is with the realization that La Mu is the same spirit that established Buddhism in India 14,000 years later, when he was reincarnated as Shakyamuni, that it becomes obvious that the teachings of Buddhism actually have their roots right back in the La Mu era.

La Mu's teachings, 17,000 years ago, represent the beginnings of true religion. In those days, religion and politics were not separate. The aim of religion was perceived also as the aim of political government, and the foremost religious leader was also regarded as the best politician. This actually makes a lot of sense, for if all humankind comes originally from God it should naturally follow that the one to lead them while they live on Earth should be one who is closest to God—in other words, a great religious leader.

Every night, La Mu would kneel in the temple and converse with the high spirits, asking them for help in questions of national policy. This is the true basis of politics, because a politician has authority over the people and if he makes a mistake it affects not only him but also the lives and souls of the whole population. For a mortal to take such power to himself is the height of arrogance. So, the first thing a politician should do is to humble himself before God, to open his heart and listen to what God has to say.

Eventually, however, La Mu died, and his great teachings were gradually eroded and forgotten, with the result that the end of Mu's golden age became inevitable. People began to deny the power of enlightenment, and a sinister worship of animal spirits spread, a perverted worship of base spiritual power that scorned the teachings of love and mercy and caused dark clouds of evil conceptual energy to spread across the whole continent.

15,300 years ago, Mu began to sink beneath the ocean in three stages. Even the great, modern city that had been named after La Mu duly disappeared beneath the waves of the Pacific Ocean. Again some of the inhabitants were able to escape in ships. Those that headed north became the ancestors of the Vietnamese, Japanese, and Chinese; others went east across the Pacific Ocean to live in the Andes. A third group journeyed on in search of a new world and eventually found their way to Atlantis in the Atlantic Ocean.

7. The Civilization of Atlantis

The era of Atlantis is the one that immediately precedes our own. Atlantis itself was situated in the Atlantic Ocean in the area that is known today as the Bermuda Triangle. Originally a small island no larger than the British Isles, it grew to form a complete new continental landmass following a submarine eruption 75,000 years ago. People first started to live there about 42,000 years ago, but the first inhabitants were a very simple race who had crossed over from neighboring islands.

Real civilization did not start on the continent until approximately 16,000 years ago, a few hundred years before Mu disappeared. It was at this time that Koot Hoomi—later to be reincarnated in Greece as the great scientist Archimedes—was born, to bring civilization to a people who until then had subsisted on fishing or hunting.

Koot Hoomi was fascinated by the mysterious power of plant life. He wanted to know why seeds should sprout, put forth stems and leaves, and produce flowers; he wanted to understand how a stem could grow from a bulb. For twenty years he studied the subject—and eventually he discovered the true energy that is the life force. As he might have said, then:

> *Life is itself a treasure-house of energy. As it changes its form, a huge transformation of energy takes place. If we could only divert some of the force of this energy, it could be used to drive many machines.*

He spent a further ten years striving to find a way to tap into the energy of the life force. Eventually he succeeded, and his innovation was to provide the form of energy that powered this civilization. A new type of light appeared on Atlantis.

The power produced by the life force was used in much the same way as we use electricity today; many "electrical" appliances were devised to exploit it. On the window ledge of each house stood a row of flasks containing flower-bulbs, and each bulb was connected by a Nichrome wire to a machine that drew off the energy it produced as it began to germinate and grow. This was fed into a condenser-amplifier where it was increased and from where it supplied all the power that the household required.

This style of life in Atlantis was dramatically altered, however, 15,300 years ago when some of the survivors from the vanished continent of Mu finally arrived there. Among them were some scientists who knew the secret of Mu's pyramid power, and they offered it to the inhabitants of their new homeland.

At roughly the same time, the Grand Nyorai Maitrayer took on human form and was born in this world under the name of Kuzanus. The message he preached was a form of Deism combining pyramid power and sun worship. He proclaimed that all things that could be explained rationally and scientifically were in accordance with God's will, for God loves that which is rational and scientific. The best example of this, Kuzanus said, was sunlight, and the light of the sun lay at the core of what he taught. "What a wonderful thing the light of the sun is," he said. "Not only can it be converted scientifically through pyramid power to benefit people but at the same time it shows us the way God's mind works." Eventually, pyramid power came to be used as the standard technology in aviation and shipping.

The culture of Atlantis came to a peak approximately 12,000 years ago under the leadership of Thoth "the Omniscient and Omnipotent." Thoth was a genius in many fields including religion, politics, philosophy, science, and the arts, and was responsible almost single-handedly for creating the final overall culture of Atlantis. He was particularly gifted in the sciences, and as a result Atlantis was able to surpass even Mu technologically, manufacturing airships and submarines that were driven by pyramid power.

The airships of Atlantis would have seemed a very strange shape to us, resembling nothing so much as a whale thirty meters (just over ninety-eight feet) in length and four meters (thirteen feet) in diameter. The upper half of the craft was filled with gas to give it lift, while the lower half was for passengers. Able to carry twenty people, each airship had what looked like three dorsal fins on its roof that were really three silver pyramids that provided the power for the propeller fitted at the rear. Because of their source of motive power the airships of Atlantis could fly only on sunny days: all services were cancelled when it rained.

The submarines were made of special alloys and shaped to

resemble an orca, or killer whale, twenty meters (nearly sixty-six feet) long and four meters (thirteen feet) wide. The orca was the symbol of Atlantis, and although some theorists claim that the nation was named after a king called Atlas, the word *atlantis* really means "the golden orca." The submarines were also fitted with three pyramids on their upper surface, which, from a distance, resembled the dorsal fins of the orca. The ships would surface to recharge their batteries before diving once more.

Atlantis inaugurated the age in which science was to predominate. After the great Thoth died, the people carried on his teachings as best they could but, unfortunately, no leaders emerged who were able to equal Thoth's vast and multifaceted knowledge. In time, society began to regard science as the be-all and end-all of everything. A few people nonetheless queried what was going on, saying pointedly that "For science to rule supreme may not be God's will after all," or "God's purpose may in fact lie in some other direction altogether." A number of minor religious reformers appeared in this way, and this period—which lasted for a thousand years—came to be known accordingly as the Age of the Hundred Debates.

This was about 11,000 years ago, around the time that Atlantis began to sink. First, the eastern third of the continent disappeared. Then, 10,700 years ago, the western third also vanished. Only the central third of the continent in due course remained above the ocean's surface, but life for the inhabitants of the civilization went on just as it always had.

8. The Age of Agasha

Some 10,400 years ago (or 8400 BC), Agasha was born in the continent's capital city, Pontis. Pontis had a contemporary population of around 700,000 and was the hereditary seat of the ruling Amanda dynasty.

Agasha was born a prince of the Amanda clan, and the name given to him in childhood was Amon. Crowned king at the age of twenty-four, when he changed his name to Agasha (which means "he who treasures wisdom"). He was in fact an earlier incarnation of Jesus Christ.

Like La Mu, Agasha was both a religious and a political leader. The temple that belonged to his palace was constructed in the shape of a golden pyramid thirty meters (just over ninety-eight feet) tall, and it was there that the king performed his religious duties. A unique aspect of his reign was that once every month he would assemble more than 100,000 people in the huge city square and preach to them. (The technology of Atlantis had naturally by this time come up with a radio microphone and public address system much like the type we enjoy today.)

Agasha's teachings were similar to those he was to proclaim when he was later reincarnated as Jesus, and were based on love. What he said was different each time he spoke, but his basic message remained the same, and may be summarized as four specific affirmations.

God is love, and as children of God all people have love within their hearts.

To practice love, first love God, then love your neighbor who is a part of God, and finally love God's servant—that is, yourself.

Every day, pray quietly on your own and converse with your guardian and guiding spirits.

Human love is measured by the quality, not the quantity, of love you give—so improve the quality of your love.

Agasha's teachings were admirable and the people in general revered him greatly, but there was one group who still held firmly to the form of Deism preached by the holy man Kuzanus (the Grand Nyorai Maitrayer), and they plotted to kill him. Kuzanus had taught that God was a rational being who assigned great importance to science and logic. Here was Agasha, on the other hand, preaching love and speaking of guardian and guiding spirits, all of which was the antithesis of science, rationality, and logic. They were convinced that Agasha's teachings would lead the people astray and put an end to the ancient traditions of Atlantis.

It has to be said that, although Agasha was a paragon of humankind such that his inherent goodness was evident to all and sundry, the average Atlantean accepted the supremacy of science and found it difficult to believe in spirits that he or she could not see. Eventually, the Deists staged a coup and buried Agasha and his whole family alive under the city square. In this respect nothing has changed; when we try to preach the

Truth today, we are opposed at every turn by the forces of evil.

When the palace was stormed, a single member of the royal family—Agasha's first son, Amon II—managed to escape in an airship and fled to Egypt. There he taught the people to worship the sun and is remembered to this day as Amon Ra. His teachings were also responsible for the initial construction of the Egyptian pyramids.

Many angels of light who were contemporarily living on Earth in human form were executed during the revolution, and it seemed for a time that evil had triumphed in Atlantis. But as the dark clouds of evil built up over the land, the Earth Consciousness reacted violently, causing what remained of the whole continent to sink beneath the sea in a single day.

Not by any means for the first time, a whole civilization vanished overnight as the landmass plunged beneath the waves. Nonetheless, some of the inhabitants succeeded in escaping in airships, some traveling to Africa, some to Spain, and some to the Andes ranges in South America, where they planted the seeds of new civilizations.

9. The Dissemination of Modern Civilization
After the destruction of Atlantis, civilization sprang up in various guises all across the globe. Amon II went to Egypt, where he was worshipped as a deity and preached the religion of light. Assembling together the inhabitants of the region, all of whom subsisted on farming, he taught them some of the won-

ders of civilization. The pyramids for which their culture in time became so famous were based on one that Amon Ra had built for his own personal worship.

About 4,000 years ago, the spirit of Jesus Christ was born in Egypt where he was known as Clario. He preached to the people, and his message combined both the elements of love and sun worship.

In South America, the descendants of refugees from both Mu and Atlantis worked together to create their own unique civilization. They believed that God was a space-traveling astronaut, and their culture was founded on the desire to communicate with Him. For this purpose they constructed enormous landing-sites for spacecraft high up in the mountains of the Andes. There came a time, however, 7,000 years ago, when King Rient Arl Croud lived and ruled in the ancient Inca kingdom in the Andean mountains. He declared to his people that God was not a space-traveler but that He lived within all their hearts. It should be everyone's aim, he said, to explore the mysteries of the heart and to try to raise themselves up and become as close as possible to God.

Rient Arl Croud was in reality a reincarnation of both La Mu from the Mu civilization and Thoth of Atlantis. He was destined to be born again later in India as Gautama Siddhartha, otherwise known as the Shakyamuni Buddha, where he spread the teachings of Buddhism. Unlike the spirits of the fourth or fifth dimension, those of the ninth dimension are a single vast entity of light energy and are not reborn as the same person each time; rather just one facet of their life-force comes down to Earth.

Between 3,800 and 3,700 years ago, Zeus appeared in Greece. Long afterward, he came to be known as "the Omniscient and Omnipotent" in recognition of his mastery both of learning and of the arts. Because he is in charge of every aspect of art in the ninth dimension, it is hardly surprising that the culture he founded should have been so full of beauty. His teachings were aimed at achieving the liberation of the human spirit. Unhappy with the tendency of human religions to make people suffer from guilt, Zeus wanted people to remain free to express their humanity—and many of the gods in Greek mythology, therefore, appear to be bright and happy individuals.

Moses was born in Egypt between 3,300 and 3,200 years ago, the son of two slaves. Set adrift on the Nile as an unwanted baby in a raft of reeds, he was lucky to be discovered and raised in the ruler's palace. He did not learn about his lowly origins until he reached manhood, but having done so he led hundreds of thousands of his people out of Egypt and across the Red Sea in search of the Promised Land. During his long career he received considerable guidance from God, including the celebrated Ten Commandments.

Some 2,000 years ago, Jesus Christ was born of the same race that Moses had rescued, grew to manhood, and preached his teachings of love. He was crucified, but rose again and showed himself to his disciples. Of course, after his resurrection he appeared only in his spiritual form, but he sat down to dinner with his disciples to prove to them that he had genuinely returned. That he had not actually been resurrected in the flesh is obvious from the way he ascended to heaven afterward.

While on Earth, Jesus had been guided by many different spirits, but his pivotal message of love and faith, and the phenomenon of his resurrection, were the work of Hermes. The reason Christianity spread to become one of the major world religions is that Jesus essentially rejected the teachings of the Jews' ancient vengeful god (Yahweh or Jehovah) in favor of the God of Love (El Cantare). It was Hermes and the other Greek gods, however, who raised Jesus from the rank of prophet to that of the means of Salvation, and who spread his teachings throughout the Roman Empire.

Some time earlier, more than 2,500 years ago, Shakyamuni preached Buddhism in India while in China Confucius preached his philosophy. The seeds of the Truth have thus been scattered throughout the world, and the result is our present disparate civilization.

10. And So to the Golden Age

When we look back over the history of the civilizations from one million years ago up to the present day, we cannot help but notice that they all share several points in common.

1. All civilizations rise and fall.
2. God (or Buddha) always provides each civilization with Great Guiding Spirits of Light.

3. As each civilization reaches its peak, evil competes with the final rays of light and causes clouds of dark conceptual energy to descend over humankind. This always results in some natural disaster—such as an alteration of the Earth's axis or the sinking of a continent—which brings the civilization to an end.

4. Each civilization inherits vestiges of the previous culture, but is based on different values.

5. No matter what the values a civilization is based on, it always remains an environment suitable for human souls to improve in, through the process of reincarnation.

Looking now at our present civilization in the light of these points, we may be brought to the realization that our civilization in the latter half of the twentieth century to the beginning of the twenty-first century bears a very strong resemblance to the Mu and Atlantean civilizations just before they were destroyed. We have the same emphasis placed on technology, the prevalence of materialist values, the disturbance of people's minds, the spread of social injustice, and the ubiquitous appearance of fraudulent religious leaders who mislead people and turn them aside from following leaders of the true faith.

Looking at the fate of past civilizations, it is surely quite obvious what is going to happen to our own sometime in the near future. Our present civilization is not confined to a single continent, it is worldwide—and so, when disaster falls, it will be on a global scale. Not only that, but all the signs seem to point to the end's coming within the next few decades.

I make this prophecy in the full knowledge that the disaster I can see quite clearly approaching will duly occur. I also know what is to be the fate of humankind, and one thing I can say quite definitely is that no matter how terrible the catastrophe that will befall the world may appear, it will not be the end of everything. As previous civilizations thrashed about in their death throes, their inhabitants believed that the disaster was absolute. But humanity always springs back to create a new paradise of hope, to build a new civilization that is full of light.

Just as an individual goes through a cycle of reincarnation, humankind as a whole may be thought of as being reborn from time to time. Civilization, too, happens in cycles: the end of one civilization means the birth of another.

This book, *The Laws of the Sun*, is written as a direct revelation from the ninth dimension because the whole world is about to be plunged into darkness, and humanity will need a light, a beacon in the darkness, to show it the way. The light that is needed is the light of Truth. *The Laws of the Sun* represents the rising sun of the Truth, the light for the next civilization.

After several decades of confusion, a new civilization will be formed later on this century from the ashes of the old. This civilization will originate in Asia. It will spread from Japan to the countries of South-East Asia, to Indonesia and eventually to Oceania. More than one of the continents that we know today will sink into the ocean's depths, while a virgin continent of Mu will rise from the Pacific and become the center of a vast new culture. Parts of Europe and the Americas will sink into the sea, but a new and even larger Atlantis will rise from the Atlantic—and it will be there that Jesus Christ will be reborn around the year 2400. Around the year 2800 Moses will be reborn on the continent of Garna that will by then have reappeared in the Indian Ocean, and he will create a galactic civilization.

Some of the people who are reading this book will undoubtedly be reincarnated on Earth in the future to listen to the teachings of Jesus or Moses. The success of these future civilizations, however, is dependent on our evoking the dawning of the Sun of Truth in Japan now, today. It was to perform this important task that you were all born to be alive at this precise moment.

Many of the people who helped spread the Truth in the time of La Mu, Agasha, Shakyamuni or Jesus Christ have been reborn in the present. Among them there are numerous Bosatsu (angels) of Light, and I know that they include some readers of this book.

Six: The Path to El Cantare

1. Open Your Eyes

You must not imagine that you exist merely to live in this world once or twice. In Chapter Five we looked back over the history of the last million years and saw how continents rose and fell in the oceans, how the various civilizations flourished and died. Do you think that all the innumerable people who lived in these lost civilizations were unrelated to you? Do you think they just sprang out of the ground?

If you do, I am afraid you are wrong. The citizens of Atlantis and Mu were none other than yourselves. Somewhere in the furthest recesses of your soul there is a treasure house of memory that contains the knowledge of your lives in tens or even hundreds of past civilizations. This is true of everyone, not just those with special psychic powers. It is your soul's memory, and it is something that everyone has equally. It is the wisdom you have acquired through countless times of reincarnation, although it has been hidden from you for the duration of your life on Earth.

What you think of as your individual self is not your real being, it is effectively no more than a rag doll. The body is simply a vehicle like a ship or a car that you have been given in which to come to Earth to enable your soul to grow spiritually. You are the captain of the ship, the driver of the car, not the ship or the car itself. I want you to wake up to your true self, the soul that controls your body. I want you to find the real you.

If you think you know everything about the world after a mere ten or twenty years at school, you are very much mistaken. But who can tell you who you really are? You will have to search for your true identity yourself.

But what does it mean "to find the real you"? It is to find out everything about your soul, and that can be achieved by searching the depths of your mind. Indeed, who but you could do that? Who but you could search your mind to find out this truth? If you cannot talk about your true self, who will be able to tell you? Enlightenment is an encounter, a confrontation, with one's true self; it is being able to tell people the truth of one's own mind. It is the power to say, "That's me."

Human souls are part of Buddha (God) and may be regarded as constituents of Buddha's artistic self-expression. Human beings are blessed with the capacity for creation and freedom of choice, but many of them unfortunately misuse this opportunity to waste their lives in self-indulgence. Eventually they forget that Buddha is their origin; they forget Buddha's will and devote their lives to self-centeredness and the appetites of the flesh. As soon as they find themselves more attached to this world than to heaven, they are totally lost. In those circumstances, when they return to the Real World they work to build a realm identical to the Earth they know, filled with the same desires and the same contentions. This realm is what we know as hell.

To know yourself is to know that you are a child of Buddha and to know His will. To open your eyes is to become aware of your spirituality and to open your heart to the Real World in the fourth dimension and above. If you are content to live as you do now, if you are satisfied with your present outlook on life and on humanity, please by all means settle back and enjoy life as you know it. If, however, you want to open your eyes to the Real World, you must start by exploring the depths of your mind because there you will find the clue that will lead you to the Land of Buddha.

2. Abandon Attachment

To know yourself you must first discard what you think is your "self"—in order to know your true self, it is necessary to cast aside your false self. In other words, to realize that you have a false self is the first step to abandoning it. The false self can be defined as having four contributory characteristics: there is the self that takes love from others; the self that does not believe in Buddha; the self that lacks the will to make spiritual progress; and the self that is full of worldly attachments.

The self that takes love from others

The first characteristic of the false self is the one that thinks only of taking love from others. Buddha (God) gave us the universe. Our souls and bodies also come from Buddha. He gave us the sun, the air, water, the land, the seas, the animals, the plants and the minerals but He asks for nothing in return.

We human beings live in a world where everything we could possibly need is provided, and yet all we can think of is how to get more. Buddha offers us his immeasurable love, so how much more do we need before we are satisfied? It is only people who do not know Buddha's love that try to steal love from others. But what is this "love" that they try so hard to get hold of? It is nothing more than worldly esteem.

What is the point of such esteem if it is based on earthly values? What merit can there be in respect based on three-dimensional materialist values? What use is it to anyone? How can it help us to raise ourselves up? Such self-centeredness just creates barriers between ourselves and others—barriers that multiply and accumulate until the whole world resembles a zoo with every person in it trapped inside a cage of his or her own making. Why can't everyone understand this? It is because we hold on to false attachments: that is why we cannot understand. As long as we persist in holding on to these attachments we will never be able to realize true happiness.

The self that does not believe in Buddha

But the people who deserve even more pity are those who do not believe in Buddha (God), who do not believe that this world was created by Buddha. They think that humans are simply the result of copulation and that every individual lives a life that is totally independent of the rest of humankind. This is the most wretched of the false selves.

Such people say they do not believe in Buddha's salvation

and insist that if they are to accept the existence of Buddha at all, they must first be shown proof. But by saying that, they are in effect putting themselves in judgment over Buddha; their conceit is of such a magnitude that they somehow believe they are in a position to sit in judgment.

But, of course, it is impossible to prove the existence of a being who has watched over humankind since before the creation of the world. The only way to obtain the sort of proof they want is to die and return to the Real World—by which time it will be too late. When they get there they will find themselves in a world of darkness and be so confused that they will be unable to prove even their own existence.

The self that lacks the will to make spiritual progress
The third false self is the one that takes no real thought for spiritual progress. In this respect, this self suffers from several insufficiencies: first, it is generally indolent, too lazy to care; second, it does not bother to devote itself to studying the Truth; third, it does not take the trouble to discern other people fairly; and finally, it prefers not to be open-minded. Buddha expects people to strive for the Truth all their lives, so those who do not make an effort cannot be considered children of Buddha.

Do you make a sincere effort every day? Do you try daily to find out more, to deepen your knowledge of the Truth? Do you recognize the true worth of other people? Are you open-minded? People who are not open-minded will never improve their lot; people who are not open-minded can never achieve true spiritual progress. To be open-minded is in itself a virtue and in accordance with Buddha's will. If you are forever contradicting others and not listening to what they have to say, it shows you are not open-minded.

The self that is full of worldly attachments
This false self is one that is full of spurious attachments to this world. To know your true self is to live every day with the mind of Buddha as your own. Once this has been achieved, you realize that this world is a place of learning only, and that eventually you must give up everything and return to the Real World. No matter how you may cling to the Earth, one day you will have to leave it for the next world.

So this world is just a stage through which we travel, and we must live each day as if it is our last, for we never know when we might die. There are no people living in heaven who are attached to this world—conversely, however, all the inhabitants of hell remain enamored of it. Never forget that for a moment.

3. Glow a Fiery Red, Like Molten Iron
To resolve to break free from all worldly attachments is one of the most fateful decisions you will ever have to make. Any decision which guarantees the possibility of happiness in eternal life is bound to be of the utmost personal significance. However, coming to that decision does not mean that there-

after you have to lead a relatively negative or backward-facing life. In fact, cutting yourself off from your former attachments will open the way to a positive and purposeful existence.

Take a look around at the people in this world and you will see that those who allow themselves to become attached to it weaken themselves by doing so. Why do they cling so tightly to their status or reputation? Why are they always comparing their annual income with others? Why are they so proud of their educational background or the corporation for which they work? Why are they forever worrying about what they look like or what other people will think? What do such attachments bring them? What good does it do them to win the admiration of the people of this world? Seen through the eyes of Buddha, who is greater than the whole universe, these human attachments are inexpressibly fleeting, futile, and insignificant.

A true life from which all earthly attachments have been thrown off glows a fiery red, like molten iron; it is the true life of a child of Buddha, a life of which Buddha himself would approve.

It is impossible for people to take status, reputation, or wealth accumulated on Earth with them when they die and return to heaven. Their eminence in the material world has no meaning whatsoever in the other world. Believe me: there are many ex-prime ministers of Japan suffering in hell. Indeed, many hundreds, many thousands of company presidents who were once the envy of everyone on Earth are now doomed to a life in the Hell of Lust, the Hell of Strife, or the Hell of

Beasts. There are even more people there who thought of nothing but money when they were on Earth, who led lives of pleasure surrounded by beautiful women, but who are now having to pay the price in hell. For their ten, twenty, maybe thirty years of enjoyment they now have to suffer for untold hundreds of years. Hell is not just something from old tales with which you can scare children. It exists, and it exists at this very moment.

For those who have fully awakened to the Truth, the souls suffering in hell can be seen as clearly as if they were goldfish in a goldfish bowl—and the thing they all share in common is that the more they are attached to this world, the worse is their suffering.

The essence of humanity lies in the mind and soul, and when we die the only thing we can take back to the other world is our mind. The mind is everything—and once people realize that this is all they can take with them, they can start to live with more resolution. If all we can take with us is our mind, then we should want our mind to be as beautiful as we can make it. But what is a beautiful mind? It is a mind that can win Buddha's praise; it is a mind filled with love. It is, in other words, a giving heart, a nurturing heart, a forgiving heart, a thankful heart. To create a heart-mind like this to take with you, to polish your mind, to raise the level of your mind, you must glow a fiery red, like molten iron.

What is the opposite of worldly attachment, then? It is love. "How can that be?" you may ask. It is because to love is to give. What distracting attachments could there be for a love that is given freely and constantly to nurture others? To free yourself of worldly attachments, therefore, you should start by giving love to others.

What have you done for the parents who brought you up? What have you done for your brothers and sisters? Have you lived up to your teacher's expectations? What have you done for your friends? What have you done for the people whose paths have crossed your own during your lifetime? What have you done for your lover, your wife, or your husband? As you brought up your children, did you think back over the sacrifices your own parents made when raising you? Are you reconciled now with anyone you once hated or resented? Have you helped calm the anger of others? How much have you been able to respond to Buddha's love as you travel purposefully through life?

4. Life Is a Daily Challenge

Having sworn to leave behind your worldly attachments, bared your heart, and pledged to live a life as a child of Buddha, what is the next step?

It is not to live like a hermit in the mountains, to meditate under waterfalls, to fast, or to spend most of your time in contemplative study. We are not born on Earth to hide in the mountains any more than we are born to fast. Shakyamuni proved that enlightenment could not be achieved this way as far back as 2,500 years ago, and the example of his life should be sufficient to prove to anyone that enlightenment is not to be found through deliberate, ascetic discipline.

At the same time, enlightenment is not to be found in a life devoted to sensual pleasure any more than it is to be found in one of extreme physical hardship. To live the life that Buddha expects of us, we must abandon both extremes and lead a life devoted to the Middle Way. The essence of humanity may lie in the mind and the soul, but this does not mean that the needs of the body are to be either ignored or over-indulged. Your body, after all, is a valuable gift from Buddha, created through the meeting of your parents, and provided as a vehicle for your spiritual learning.

Anyone can acquire a car: all it takes is money. Some people, though, become almost obsessive about cars and spend all their free time working on them. If they can look after a mere machine in that way, think how much more care we should take of our bodies. To stay healthy we should exercise, eat a balanced diet, make sure we get enough sleep, and live a regular life. We should also be careful that we do not lose our soul to alcohol, which can distort our powers of perception and reasoning. Once people feel they cannot live without alcohol they stop thinking clearly and open themselves to the devils of hell. The result is always an eventual failure in the workplace and the destruction of the family.

Of course, it is easy to say that we should ignore both extremes and take the Middle Way—but it is very difficult to achieve this in actual fact. The more we think about the

Middle Way, the more profound it becomes. It is not long before we find ourselves asking exactly *how* we can live our lives in the Middle Way. What can we use as a standard against which to measure our lives?

There are two standards available to us that can lead us to the Middle Way. One comprises the Eightfold Path and the life of meditation upon the self that it encompasses. The other is based upon the stages of the development of love (see Chapter Three) and the contemplative betterment of self that they embody. Our lives should be grounded in these two standards.

The Noble Eightfold Path, as I have mentioned, teaches Right View, Right Thought, Right Speech, Right Action, Right Livelihood, Right Effort, Right Mindfulness, and Right Concentration. Based thus on a standard of "rightness," it provides a method of preventing your thoughts and actions from swinging to either of the extremes, and so helps you to find the Middle Way. By taking the Middle Way you will be enabled to live at peace with your fellow creatures, to live a life of genuine and perfect harmony. Nonetheless, you must take great care not to allow yourself to become so involved in trying to live a life based on "rightness" that you end up with a disposition that is simply negative or pessimistic. Put too much into your self-examination and you may find it difficult to progress any further.

Once you are able to reflect upon your self to an appropriate degree, the next task is to express the results in your thoughts and your actions. One way of achieving this is by means of exercises in gratitude. What "exercises in gratitude" do I have in mind? Well, one way of showing gratitude is of course by saying "Thank you." But real gratitude must be expressed in a more positive way, by asking yourself what you can do to help others. This is love. It is an act of giving; an act of continually giving. It is the practice of giving love without expecting anything in return. This is the way to be truly grateful. From time to time you must take a look at yourself and decide whether you have reached the stage of development in love (see Chapter Three) that we have called fundamental love, or whether you have reached forgiving love, or whether you may even have reached the level of love incarnate. The stages of love in this respect provide you with a measure of your progress and enable you to gauge your development day by day.

Think back and consider whether or not you are making progress. This will tell you if you are living the life of a true child of Buddha. Think every day of the occasions you have thought or done something extreme, and in the light of those occasions consider what progress you might have made for the day. Only by doing this will you truly be able to claim that you are overcoming the challenges of everyday life.

5. A Sparkling Life

Certainly, reflection upon the self and measurement of progress are essential. But I must say that such considerations by themselves make for something of a colorless existence.

Life needs a bit of sparkle. How can we put a bit of sparkle into our lives?

The sparkle I mean occurs at the moment when brilliant light is emitted from the soul, a phenomenon that I believe arises as a result of any one of three different causes.

The first is the curing of a sickness. Illness is a time of ordeal in our lives and our humanity is tested through the way we choose to contend with it. The ordeal that illness represents is twofold; we undergo both physical suffering and an affliction of the spirit.

The physical suffering is commonly the result of an irregular lifestyle, overwork, or mental stress on the part of the patient. It is important to realize that there is a significant mental element to physical illnesses. Indeed, the root cause of some recognized disorders lies entirely in the mind. When you fall ill, therefore, be prepared to consider mental stress as at least one possible cause.

Nonetheless, around eighty percent of all human illness is affected by the possession of spirits, which are drawn to the negativity in a person's mind. The spirits of the dead take over a person's body, causing it to share their sweating and suffering. As the spirits are exorcised, it becomes evident that this is the case: the sick person's fever disappears, his or her body feels refreshed, and he or she is able to move around again. It just shows how susceptible the human body is to spiritual influences.

Above all else, these spirits of possession detest self-reflection and gratitude because when a sick person begins to look back on his or her thoughts and deeds, repent and feel gratitude, an aura begins to extend from the back of his or her head at a wavelength incompatible with that of the spirit, rendering it impossible for the spirit to remain in possession any longer.

To make this aura shine even more brightly, the sick person must eliminate the source of any spiritual anguish. The way to do this is to look closely at whatever ties him or her to this world—his or her worldly attachments—and, one after another, lose them.

Describing the treatment of the sick in this way may sound prosaic and may seem to have little to do with medical science. But, if you are ill, once you have eliminated all your worldly attachments and feel that you are ready to die at any time, your guardian and guiding spirits will cause the light of Truth to flow through you—and the illness will rapidly disappear. This is a true miracle. People who go through this moment of miraculous healing experience something of a religious awakening, which dramatically alters their lives. The light that then sparkles affects not only the sick person but shines also into the hearts of those around them.

The second moment of sparkle in our lives is when we awaken to faith. The difference between a life without faith and one that knows Buddha is the difference between a man groping through the dark and one who carries a bright lantern to light his way.

The third dimension is a very materialist world, and if you

allow yourself to become enamored of worldly goods to the point of believing that materialism represents the whole truth, you will gradually abandon yourself to pleasure or strife. You will forget the eyes of Buddha and become the most wretched type of human being. Faith is a beam of light to show you your way. It is only through this light that we who are unable to see the Real World can have our eyes opened.

The third occasion on which this sparkle comes into our lives is when we experience the phenomenon of a "spiritual message"—when our spiritual heart opens and we are able to hear the words of our guardian or guiding spirit in our subconscious. This phenomenon is something we talk a lot about in the Institute for Research in Human Happiness. It is a communing with high spirits that can be achieved only by those who have attained enlightenment, and is characterized by the fact that, through it, the spirits impart to us important new teachings about the Truth. It means that while we are learning from the high spirits, we are also under their protection. It means, in addition, that as our knowledge of the Truth grows, more and more opportunities become available to us to repeat the experience. Both direct and indirect contact with the high spirits in this way can lead us to experience the third type of sparkle.

6. Time Is as Precious as a Diamond

To make full use of the powers we have been given in this life we must also make full use of our limited time-span. A soul is only given the opportunity to be reborn on Earth once every few hundred years, or even once every few thousand years: it is such a valuable experience. Despite this, the vast majority of people live their lives without ever giving any thought to why they are alive: they waste precious time. Even if they awaken to the existence of Buddha in their final years and pledge to make a new start, they can never make up for the years they have lost. Time is like an arrow once loosed, like water in a river: once gone, it can never be regained.

For this reason, those who awaken to the Truth in their youth are doubly blessed—and if they can live their whole lives in Truth, they are the most fortunate of all. This is not by any means to suggest that all is lost if one does not realize the Truth until late in life. For, if from that point, one lives the remainder of one's life to the full in the service of Truth, one may yet be held to have lived a valuable life.

There is a secret to living a full life. It is to imagine how you will feel when the time comes for you to die. What will you think? How will you feel? If after considering these questions carefully you think you will be able to say, "I am glad I lived," or "It was a good life," you will have lived a happy life. If, on the other hand, you reckon that you will be full of regrets, then you are to be pitied. When we return to the Real World, we will be obliged to look back over our lives in front of the assembled high spirits. Our whole life will appear before our eyes as if we were watching a vast cinema screen, and there will be countless other people present to watch.

In this way the people who have just died learn what kind of persons they were in Buddha's eyes. The time for lies or excuses is long past—feeling the weight of a multitude gazing on them, they realize where they belong. People who belong in hell make the decision to go there themselves: on realizing what kind of person they have been they are too ashamed to want to live in heaven. In slightly more technical terms, people whose spiritual wavelength has been coarsened by their lives on Earth find it impossible to blend with the refined vibrations of the inhabitants of heaven. They have taken on so much of the materialism of this world that their spiritual bodies have become cumbrously heavy, and they naturally sink downward.

Some, however, are hailed as heroes when their lives are shown before the assembled spirits. When they come to the scene where they have become aware of Buddha and clasp their hands in supplication, tears streaming from their eyes, they are given a thunderous ovation; all the other spirits come over to clap them on the shoulder or to shake their hand. When they come to the moment at which they stand up and swear to devote their lives to the spreading of the Truth, even the Bosatsu of Light cannot withhold the tears.

This scene awaits us all, although the number of years before it happens is individual to each one of us. Death is one fate from which there is no escape, and you should think about it and be prepared for it. In fact, you should ask yourself what would happen to you if you were to die tomorrow. Would you be proud of the way you have lived? Would you have any regrets? Look to your conscience for guidance on your answer. This knowledge and change in outlook is necessary if your time on Earth is to shine and sparkle like a diamond.

So imagine that you are about to die. Look back over everything you have ever done in your life. Think of it as if some benevolent third party were reviewing your life from start to finish. This is the secret to living a life that sparkles with the translucent luster of a diamond.

7. Embrace a Dream

Everyone needs to dream. A life without dreams is a life without hope. Of course, it is important to be sorry for the wrongs you have committed and to regain your integrity—but this will only take you from a negative account up to zero. A life without a plus is not enough.

To embrace a dream is to plot the best possible course for your life. For instance, when you have a house built you employ an architect to draw up a plan and the builders follow the plan to create a wonderful house. In terms of life, however, the architect is none other than you yourself, and if you do not draw up a proper plan the result will be a haphazard affair. To build a house you might take a lot of trouble over the planning—so you should take at least as much effort over structuring your life. Too many people live a reckless life just taking everything as it comes.

Not much is needed in the way of a plan, after all. You

simply have to embrace a dream, to have a dream of the future. You can tell who does have a dream and who does not by the confidence they show, by their powers of persuasion. When you meet someone with a dream, it fills you with joy; it inspires you to do something about your own life while wanting to help them in theirs.

There is something about embracing a dream that intoxicates people. I am sure it is true that every person who has performed any great feat on Earth started with a personal dream. Having been born on this world, it is important for us to have the enthusiasm and energy to try to accomplish something great. Modesty need not prevent you from thinking big. Modesty is a necessary part of growth, but it is really needed only when you become overconfident. In other words, modesty acts as a kind of brake, but you cannot drive a car with only a brake pedal. To make the car move forward you need an accelerator, for without an accelerator the car will not fulfill its intended purpose. The brake is there as a mere precaution; it is there simply to stop the car from running out of control and crashing.

On quite a few occasions in this book I have warned about the danger of falling into hell. Awful danger that it is, however, you must not spend all your time worrying about it. If you spend each day praying and chanting the sutras you will never achieve any growth. You must put your foot down on the accelerator and only use the brake when you think you are traveling too fast. That is what the brake is for. If you are liv-

ing your life in a positive way, progressing toward your elect-
ed destination, you need only check now and again to make
sure that the brake is still working. So long as you are certain
that the brake will work when you need it—to redress the situation if you should ever make a mistake—you might as well
just use the accelerator and travel at full speed. That is the way
to embrace a dream and make that dream come true.

A dream is more than a mere plan for life; it also possesses an inherent mysterious power. A dream is a vision that per-
sists in your mind and communicates itself to the other world—
that is, to your guardian and guiding spirits. These spirits are all
the time striving to find ways by which to guide and protect the
people on Earth. But the thoughts of the people themselves are
like bubbles on a lake, and they have no plan for their lives, no
clear idea of what they want to do. How can the spirits guard
or guide such people? They cannot guide their every move
because that would be to deprive them of their independence of
will. All they are permitted to do is provide us with inspiration.
If we have embraced a dream, however, the guardian and guiding spirits can consider how we might go about achieving it,
and supply us with the inspiration to do it.

So, if we have a clearly defined dream, the spirits in the
other world may be able to help us bring it to fruition. This is
the true meaning of self-realization. First we have a dream,
which is converted to a vision; then we pray to the guardian
and guiding spirits for help and it is eventually brought to reality. It goes without saying, of course, that your dream should

be one that leads to the improvement of your soul or the spread of human happiness.

8. Take Courage

Courage: the very sound of the word makes my pulse quicken, and I wonder if I am alone in that. When I hear the word, it makes me think of an axe being driven into a huge tree. I can almost hear the steady stroke of the axe sounding through the morning forest like the pulse of life. I believe it is because they have the axe called courage that people are able to cut their way through the giant trees of adversity they meet in the course of their lives.

When you feel you are about to be crushed by the pressures of life, remember the axe of courage. In weakness or despair, remember that Buddha has provided you with the axe of courage.

When we are born as humans, we become blind. We only have our five senses to guide us as we grope our way through life, and for this reason Buddha has entrusted each of us with the axe of courage and told us to cut our way through the forest of fate. We all have this axe hanging from our belt—only all too often we never think to use it. When we are in trouble we should not turn to others to get us out of it. When we are sad we should not cry on others' shoulders and expect sympathy. Instead we should take the axe of courage and cut our way through the ropes of fate that bind us.

Let me now quote a famous Zen koan called "The Mighty

Giant" from a book entitled *Path Without a Gate* by the famous Chinese priest Hui-kai (1183–1260). It is the twentieth story in a collection of forty-eight.

> *People have forgotten the immense power they carry within them. They have been hypnotized by "common sense," by what people would think of them and by the words of doctors. They believe that they are nothing more than a physical body that could collapse at any moment. The real truth is, however, that you are a child of Buddha and as such, you wield infinite power.*
>
> *When you have found liberation through meditation, you will realize your real stature as a giant who can look down upon this world. The three-dimensional galaxy is no more than a puddle to the higher dimensions. The realms of human enlightenment below the Nyorai and Bosatsu realms appear so tiny that you must bend down to see them.*

The priest Hui-kai had evidently attained the enlightenment of the Nyorai Realm. Once this level has been reached, it becomes clear that humanity's true shape is not a soul that can be sealed up in a body somewhere around two meters (five or six feet) tall, but a force of energy that fills the universe. Indeed, this can sometimes be experienced when meditating; your body seems to grow until you can look down upon the world.

Every human being begins as a giant of awesome power, with total freedom and versatility. But we become bound by the three-dimensional senses, the "common sense" that is drummed into us at school and in society. Eventually we come to believe as a fact that there is no such thing as spirits or the other world, and voluntarily bind our hands and feet. When we are sick, we cry out that we do not want to die, and become tiny, pitiable creatures.

You must use your courage: swing the golden axe and fell the lofty tree of delusion. Be valiant—swing that axe with all your strength to conquer worry, banish pain, and free yourself from the ropes of fate.

Courage is something that is invaluable to have. When you summon it to you, you realize that you are indeed that giant of awesome power. But even when you have found this courage, have risen from your sickbed and started to live a vigorous life, even when you have freed yourself from the delusions of materialism and are aware of the Truth, you will still remain exposed to the vibrations of the materialist world. If you give in to the temptations that face people in the third dimension, your strength will soon start to wane.

If this happens, you must clench your teeth and fight on. Running a marathon, you may reach a stage at which you feel so exhausted that you cannot take another step—but if you persevere, the moment passes. If you stop, you will lose your chance to win. It is a race and the aim is to finish. If you keep going, you become aware that your feet feel lighter and that you are able after all to make it to the end. I am sure that many of you must have experienced this yourselves. It is the same when swimming a long distance; your breath becomes strained and your whole body seems to cry out for you to stop. But if you do not give in, if you keep swimming, your body becomes one with the water and you are able to plough on forward like the crest of a wave.

Life is not the same as a marathon or a long swim, of course. But in many ways it can be said to be very similar. When circumstances get rough you are expected to bear up under them, and when you succeed you gain a new confidence and can feel the light of Buddha.

9. How It All Started

Let me begin my own story as a teenager, an age at which—in my final years of elementary schooling—I was able to put in particularly long hours of study. Accordingly, my average grade for the sixth year at the school was 99.7 percent, and I passed the entrance exam for Kawashima Junior High School with top marks. I was duly chosen to represent the new intake of students at the enrolment ceremony. Looking back now at that time, it is with happy memories that I recall being head of the Students' Union, captain of the tennis team, and reporter, editor, and publisher of the school journal.

My success in academic studies continued thereafter (I managed to come first in the national examinations on several occasions), and with no detriment to my personal life. Indeed, while I was in my third year, a teacher remarked to me that it was unusual for someone who was top in his year to be genuinely popular with his peers, as I was. He went on to say, "For some reason people seem to listen to you, no one ever seems to disagree with what you have to say." I can remember it as if it was yesterday, and I realize now that if what he said was true it must have been because I already possessed some of the powers that were to lead to my becoming a religious leader.

When it was time for me to go to senior high school I decided to apply for a place at the Tokushima Jonan Senior High, the best school in my prefecture. It was just at this time that there was considerable debate about how students who went to high schools in the cities tended never to return to their hometowns again afterward. A new entrance examination scheme had been, therefore, put into operation, following the Tokyo model, by which the top ten percent were permitted to attend the school of their choice, and the remaining ninety percent had to go to local schools as allocated by a lottery system. I intensely disliked the idea that my future might be decided by a lottery system, so I studied feverishly day and night, and in due course was top student from the non-urban area and entitled thus to attend Tokushima Jonan. Now this was a school that had the excellent record of placing more than ten students a year at the University of Tokyo, and I was determined in my turn to be one of their number.

My time at the senior high school, meanwhile, was unfortunately nothing like as happy as my days at junior high had been. I did enjoy participating in the kendo club, with which I became fairly closely involved, but the two-and-a-half-hour trip to and from the school meant that I was constantly tired. Nearly all of my English studies were carried out in the dim light inside the noisy train compartments. I used to stand in the swaying carriage, textbook in my right hand, dictionary in my left, a fountain pen between my fingers, and a painfully intent expression on my face as I struggled to answer the English questions. One day, a little girl aged perhaps around four felt so sorry for me that she stood up and offered me her seat. Rarely have I ever been so embarrassed.

But although it seemed to me that I never had enough time to study in the way I wanted, I nonetheless managed again to come top in my class. My favorite subject was Japanese, and in the first year at Tokushima Jonan I got top marks in a national correspondence course no fewer than six times. This gave me a lot of self-confidence, which has stood me in good stead in later years as a writer and public speaker.

Other subjects I particularly enjoyed were earth sciences (including geography) and biology. I found I also had some flair for arts subjects, but it was on a science course that I finally resolved to concentrate for my second year—specifically in mathematics and physics, in an endeavor to make up for my shortcomings in these fields.

My first and my second years were enlivened by my being selected as the male lead in the school play. When much later I came to talk in front of an audience of tens of thousands, I often felt that I should have made rather more of my youthful stage experiences. By then, of course, it was too late.

For my third and final year at Tokushima Jonan I went back to the arts class. Standards were exceptionally high (five of the class later graduated to the Law Department of the University of Tokyo, and one to the Economics Department). Not entirely satisfied with my own academic achievements, I was happy to be awarded a prize for overall diligence and, in the spring of 1976, I finally entered the Department of Law.

Now in Japan, those who study at the Department of Law have the reputation of being the academic cream of the country, and I was worried I might not be good enough. Feeling desperately in need of extra learning, I immersed myself in a variety of associated subjects: law (of course), politics and political theory, social studies, history, philosophy, economics, commercial and industrial management, physics and chemistry, and international relations. I read textbooks in English and in German, and was surprised to find that I could read and take in English much faster than the university lecturers and professors. I remember once turning up at my local coffee shop with a thick tome on political history in Europe, and becoming so involved in reading it that when I eventually looked up it was to find that the owner was quietly diverting all his other customers to other parts of the coffee shop so I would not be disturbed.

But my time was not all spent studying. I liked to walk around the local park of an evening, for instance, and then continue to amble from there to Umegaoka, scribbling odd verses of poetry as they occurred to me. I used to look toward the western sky as dusk fell and muse about Plato's thoughts on the spirit world, or Kitaro Nishida's *Pure Experience* or *Looking Into One's True Nature.* Unconscious though I was, in those days the initial stage of spiritual awakening was in progress within me in order that I might later become a religious leader.

My happy days on the Liberal Arts campus in Komaba came to an end, and I moved on to the Law Department at Hongo. Among other projects that were well received and gained me some academic kudos at this time, was a research paper I wrote during the spring vacation of my third year entitled *The Values of Hannah Arendt.* Arendt was a US political philosopher strongly influenced by ancient Greek political thought, and was notorious among my fellow-students for writing in excruciatingly difficult Germanic English. I read all her works and sat up toiling away on my paper until six in the morning every day for two weeks. At the end of it, my friends were not impressed by the result of my labors; they declared it as difficult to comprehend as its subject was to read. My teacher, however, pronounced it a work of maturity, and said it showed promise of great things to come in the academic world. Expanded, he said, and given a proper introductory section, it might even be suitable for submission as a doctoral thesis. At the same time, my teacher admitted to wondering in the light of it whether I was really cut out for law school. Anyone who could write with such empathy about philosophy, he thought, might not be able to tolerate the necessarily rigid pragmatism of law.

I was by this time still only twenty-one, and at what seemed to be the beginning of a career in law and/or politics. Yet I tended to make light of such practical subjects as constitutional, civil and criminal law, preferring instead to focus my interest on abstract metaphysical subjects. My teacher, who retained high hopes for me, duly felt obliged to stress the importance of being grounded in more down-to-earth subjects—and so I joined the mass of students in the library at Hongo studying the Japanese legal code and all the cases and precedents that it comprises. But I remained unable to accept law as a genuinely academic subject of study in that sense.

In relation to constitutional law, for example, I could not be entirely happy about the overall intention behind its inception. I felt sorry to see my fellow students carefully memorizing its introduction and directives. As for criminal law, I could not understand how people might imagine they had the right to judge others; nor could I fathom what criteria should be used to define crime. I was certainly less than convinced by the explanation given in the standard textbook, *Introduction to Criminal Law*. Civil law seemed to bear little or no relation to Hegel's *Philosophy of Law* (also known in English as the *Philosophy of Right*). Commercial law was unutterably boring—nothing could be as sleep-inducing as those case-studies of company law. My disposition was much more toward the philosophical.

I had similar reservations against the contemporary teaching of political studies. One of my professor's attempts to include Kunio Yanagida's work on ethnology and Shichihei Yamamoto's theory of Japanese origins in a lecture on "Political Processes" did nothing but fill me with disappointed disgust at its illogicality. I certainly took a mild interest in international politics generally, but was again disappointed to hear a university lecturer lambast the US–Japan Security Pact solely on account of his own left-wing sympathies. His argument was well expressed, but his fundamental premise was plain wrong, and was indeed ten years thereafter proved to be wrong when the Cold War ended, the Soviet Union collapsed, and my own intuition was justified.

I came to realize, then, that I was disillusioned about learning law and politics at university, and that anyway there was not a single teacher in the University of Tokyo's Law Department under whom I wished to study. I would consequently have to strike out on my own. I would have to find some way of accumulating enough money to allow me the time either to cast around until I discovered an academic subject worth my studying, or, failing that, to create a completely new field of my own.

At the beginning of my fourth university year, therefore, I regarded my preparations for the Bar exams as an integral part of my search for a future vocation. I attended a specialist school for six months, where I did well enough to see one of my dissertations selected as a model answer for future classes.

But, although I passed the question-and-answer part of my exams with flying colors, my thesis was deemed not to be up to standard. My own views and opinions were by now firmly established in my mind, enabling me without reservation to point out anomalies in some of the conventional legal theories and to dispute some of the accepted interpretations of case precedents. It probably also appalled the people marking my paper that I went as far as to criticize the country's Supreme Court.

When I later discussed my failure with the high spirits, however, I found that they had been resolved that I should not pass the examination. They had in fact been prepared to take any measure they thought necessary to prevent me from adopting an ordinary career and lowering my sights to mere earthly success. It was my destiny to be a religious leader: I would in no circumstances have been allowed to pass that exam.

The personnel manager of a banking corporation linked with the national government then came to see me to offer me a job. Trying to reconcile me to my exam result he declared that no fewer than half of all the students of the University of Tokyo who passed the question-and-answer part of the examination first time then went on to pass the whole exam later—a ratio of second-time triumph much higher than at any other university in the land. He was quite sure, he said, that my being in the prestigious Law Department and majoring in politics meant that I would pass next time round, and with top marks.

But, by now, I had no wish to make my livelihood in the stifling cocoon of lifelong security represented by the civil service or its equivalent. I resolved to start from scratch, which, apart from anything else, would certainly test my personal abilities.

I was interviewed next by the personnel manager of a huge trading organization, who was insistent that he had just the job for me. I took it, much to the amazed displeasure of my well-meaning friends. One friend of mine simply could not understand how I could have turned down the offer of the job at the bank, an opening that was unique and for which my teacher himself had recommended me. Other friends were worried that I should find myself out of my depth in a trading organization, particularly in personal terms. "You don't drink, you don't play mah jong, you're not too clever at getting along with people in general, and you've never been abroad," they said. "You're definitely not cut out for life as a businessman." My closest friends were genuinely anxious for me, and as time went on they became ever more voluble in their concern.

Graduation got nearer, and as it did I recovered my enthusiasm for studying and immersed myself in such books as Carl Hilty's *On Happiness* and Heidegger's *Sein und Zeit* (*Being and Time*). A yearning to become a philosopher surfaced within me, and I spent my days reading books on philosophy and religion in an effort to discover the meaning of life.

It was at this time, in the course of just a few days, that I read all the books of Shinji Takahashi. I found that I already knew much of what he had to say, and although his work lacked the careful ordering of information that comes with an academic background—indeed, it contained many outright contradictions—I could not but sympathize with the way he strove to resolve the difficult questions surrounding the mind and how it works, and tried to fuse together religion and science. Yet despite his practical credentials (for a short time he was enrolled in the Electrical Engineering Department at Nippon University), his attempt to define and explain Buddhism in terms of basic science was not at all acceptable in my eyes. There was in addition the fact that, although he claimed to be a reincarnation of Shakyamuni (Gautama Siddhartha), he was almost totally ignorant of Buddhist philosophy. I was later to learn that my suspicions of him were well-founded.

10. The Road to Enlightenment

On the afternoon of 23 March 1981, I sat back in the warm spring sunshine and reflected on my life up to and including the period at university, and wondered what I should do in the future. The conclusion I came to was that I still wanted to establish myself as a philosopher by the age of thirty. I looked on it as my destiny.

But it was evident that I would not achieve this ambition unless I could become financially independent. In the meantime, my work at the trading organization would pay the bills and allow me to eat, and as I went on learning about society and continuing my private studies, I felt a way to do what I intended would surely open for me.

Suddenly, I sensed an invisible presence with me in the room, and almost simultaneously understood by intuition that whatever it was wished to communicate with me. I ran to get a pencil and some blank cards. My hand holding the pencil began to move as if it had a life of its own. On card after card it wrote the words "Good News," "Good News."

"Who are you?" I asked. My hand signed the name "Nikko." I was experiencing automatic writing under the control of Nikko, one of the six senior disciples of the thirteenth-century Buddhist saint Nichiren.

I was astounded. I had had no contact at all with the Nichiren sect of Buddhism. Furthermore, I was aware that "Good News" is equivalent to the word Gospel in Christian terminology, and I realized then that I had just experienced a sort of religious awakening. I was particularly astonished to have it made crystal clear to me while I was still alive that the spiritual world existed in reality, and that humans really did embody a life force that was immortal.

It was then that I grasped that my own spiritual eye had been starting to open over the previous two or three months. I had occasionally been disturbed by strange flashes of light in my eyes; once or twice I had glimpsed a golden aura extending from the back of my head. Some time earlier, when I had paid a visit as a liberal arts student to the temple complex at Mount Koya, I had had a vision as I approached the inner sanctum: I saw myself in the future working with psychic forces. That had been the year, too, that I had come

across a book by Masaharu Taniguchi called *Shinsokan*
("How to Visualize God") in a second-hand bookstore. One
night I tried out the technique that he described, but when I
put my hands together I was startled by the fiery current of
energy that immediately flowed through them, and I never
opened the book again. I thought his teaching was unoriginal
anyway.

An even earlier spiritual experience came to me in my
final year at elementary school. Lying in bed recuperating
after suffering from a very high temperature, I engaged in
astral travel several times. I visited heaven, and I also visited
the Hell of Agonizing Cries in the lowest depths of hell. So,
from an early age it was obvious that I possessed a strong pre-
disposition toward, and sensitivity for, spiritual matters.

My contact with the spirit of Nikko did not last long, but I
was soon thereafter contacted by Nichiren himself. He
instructed me to "Love, nurture, and forgive," a form of teaching that was a precursor of the doctrine of the stages in the
development of love that I was to formulate in later years (see
Chapter Three). At the time, I wondered whether I had been a
priest of the Nichiren sect in a previous life because the spirit
of Nichiren continued to visit me regularly for at least a year
afterward. Now, however, I am convinced that his purpose was
to have me refute the false teachings since propagated in his
name.

During his lifetime on Earth, Shinji Takahashi was highly
critical of Nichiren in his published works. In a speech he

gave, he went as far as to claim that when he died, Nichiren had had to spend 600 years in hell. Now Nichiren might well have been wrong in much of what he taught and much of what he did, but I learned from my spiritual communication with him that he had repented and was presently numbered among the high spirits.

In June of that year I received a message from Shinji Takahashi himself. Its contents were straightforward:

When I [Shinji Takahashi] was alive on Earth I claimed to be a reincarnation of Shakyamuni, and suggested as much in my writings. But it was not in fact the case; I was wrong, and I should like to make my apologies to everyone I misled.

You, Ryuho Okawa, are a genuine reincarnation of Shakyamuni.

Although I knew it was against the wishes of the guiding spirits, I went ahead on the advice of those around me and formed the religious organization called (in English) the God Light Association (GLA). It had been intended that I should be a psychic medium, but instead I became the leader of a new religious cult, which turned out to be the cause of considerable suffering for my wife, my daughter, and my pupils in the five years after my death. I apologize for that.

Accordingly, you, Ryuho Okawa, need feel no obligation to take responsibility for God Light Association

*as an organization or to adopt its teachings; you are
entirely free to go your own way.*

At this time I was still a businessman. I had no idea at all
of the confusion that had gripped the God Light Association
on Takahashi's death, not least that the cult was on the verge
of disintegrating altogether. So I had no conception either of
the problems that he faced. Since then, however, I have carried
out some major research in the spirit world and discovered the
whole truth for myself.

During his lifetime on Earth, Shinji Takahashi acted as a
spiritual medium, claiming that he himself was a reincarnation
of Shakyamuni and that his pupils were all reincarnations of
Buddha's ten disciples. It was not, however, the spirit of
Shakyamuni who was guiding him but that of a Sennin (hermit wizard) called Alara Kalama. (Alara Kalama was an ascetic yogi with whom Shakyamuni stayed for a short period during his six years of training; he was also the spiritual author of
the book *Shakyamuni, the Man.*) The past lives cited by
Takahashi as those of Shakyamuni's disciples were actually
past lives of Alara Kalama's followers.

In a later life, Alara Kalama was known as Zhang-xue and
lived in China during the latter part of the Han Dynasty (the
late second century), where he was the leader of a religious
movement that rose in rebellion against the contemporary government.

In the twentieth century he was born in Japan under the

name Ikki Kita, and it was this spirit who spoke through Shinji Takahashi and who had particularly harsh words to say about the unquestioning faith of Nichiren's followers, notably the Soka Gakkai cult.

Shortly before his death, Shinji Takahashi stated that he was an incarnation of an entity called El Ranty who had not visited the Earth for 300 million years. This was not true. His actual previous incarnations are listed below.

1. Enlil	ancient Sumeria	2800 BC
2. Jacob/Israel	ancient Judea	1800 BC
3. Sanat Kumara	northern India	8th C. BC
4. Zuo-ci	China	2nd/3rd C. BC
5. En-no-Ozunu	Japan	late 7th C. AD
6. Shinji Takahashi	Japan	1927–1976

Enlil (1.) is famous as the ancient Sumerian god of the wind and the sky. While in this incarnation on Earth he received guidance from El Cantare, at that time known as Anu, and worshipped as the god of the celestial firmament. Enlil was a being of the ninth dimension.

Jacob, also called Israel (2.), was the founder of the Twelve Tribes of Israel, and was a spirit of the eighth dimension.

Sanat Kumara (3.) came from the Indian Yoga Realm, and was a spirit of the ninth dimension.

Zuo-ci (4.) is renowned for having put a spell on Cao-cao, one of the heroes of the Chinese historical epic *San-guo-zhi*

("The History of the Three Kingdoms"). He was a spirit of the Chinese Sennin Realm in the seventh dimension.

En-no-Ozunu (5.), also known as En-no-Gyoja, was a Shaman who lived in the Katsuragi area of what is now the Nara prefecture, and is celebrated as the founder of Shugendo ("mountain" asceticism). He was a spirit of the eighth dimension, and his mother Hakuto-me was reincarnated as Shinji Takahashi's wife.

Shinji Takahashi (6.) dwells at present in the Japanese Sennin-Tengu Realm in the seventh dimension. (At his death he proceeded first to the sixth dimension. Publication by the Institute for Research in Human Happiness of the teachings for which he had acted as spiritual medium, however, enabled him to make his way then to the lower levels of the seventh dimension.) His daughter Keiko is a reincarnation of Akechi Mitsuhide's daughter, Hosokawa Galasia.

So these are the basic facts. Although the God Light Association effectively masqueraded as a primitive form of Buddhism, concentrating on spiritual channeling by mediums, it was in fact a Sennin cult. Paranormal phenomena (such as mind-reading) manifested as the work of Shinji Takahashi were actually brought about by such Sennin operatives as Zuo-ci or En-no-Ozunu, and it was because of this background that the group held no brief for genuine study of the mind and actually derided the evangelism of established religions. Using psychic powers to intrigue and convert people into becoming adherents is typical of Shamanists like En-no-Ozunu, and is totally contrary to the established Buddhist faith.

During his lifetime Shinji Takahashi also claimed that Shakyamuni had preached only to uneducated and illiterate people in India. In addition, he reproached the established Buddhist sects for having (he said) distorted Buddha's teachings in favor of their own or others' interpretations. In both of these respects he was wrong.

It is a historical fact that Shakyamuni was brought up as a prince. He was well educated and had a keen insight into contemporary problems both in academic learning and in social issues. For the most part, his religious community was made up of three castes: Brahmins (the priestly class), Kshatriyas (the warrior caste) and Vaisyas (the merchant caste). Fully aware, both spiritually and philosophically, Shakyamuni did not hold with differences of caste within his own faith but nonetheless respected the existing social order. To those kings and rulers who became his followers he offered advice on politics and international relations.

Because of his own personal background of youthful repression and academic poverty, Shinji Takahashi tended to associate anarchic libertarianism and contempt for social or educational superiority with the basic tenets of Buddhism. He taught that incarnations of Shakyamuni included Amoghavajra, T'ien-t'ai Chih-i, Dengyo, a certain Kukyo (supposedly an otherwise unattested Japanese priest of the fifteenth to sixteenth centuries, but probably an imaginary figure), and Kido Takayoshi. But such a list goes beyond mere religious ignorance and constitutes an insult to the spirit of Buddha. To put the record straight: Amoghavajra was a brother soul of Shariputra, one of Shakyamuni's senior disciples; T'ien-t'ai Chih-i was a disciple of Manjushri in a previous incarnation, and was the model for Sudhanashreshthidaraka ("a boy of good fortune") in the Garland Sutra; Dengyo (also known as Saicho) is presently in hell where he is working for eventual rehabilitation; an unknown priest as Kukyo could not possibly be an incarnation of Shakyamuni Buddha; and although it is true that Kido Takayoshi is a high spirit, his background is that solely of a politician.

The true earthly lives of the spirit of Shakyamuni Buddha are:

1. La Mu	Mu
2. Thoth	Atlantis
3. Rient Arl Croud	ancient Inca Empire
4. Ophealis	ancient Greece
5. Hermes	ancient Greece
6. Gautama Siddhartha	ancient India
7. Ryuho Okawa	present-day Japan

This constitutes the spiritual brotherhood of Shakyamuni, which exists in the ninth dimension. I should like to make it clear that the terrestrial group of spirits is represented by the El Cantare group, whereas the Sorcery Realm, the Yoga Realm, and the Sennin and Tengu Realms under Enlil belong

to Minor Heaven. All of my published books will be re-edited
to reflect this order in the future.

11. The Appearance of Christ and the Mission of Buddha

In June 1981, the spirit of Jesus Christ came down to tell me
something absolutely extraordinary. He spoke with a trace of
a foreign accent, but what he said was full of powerful sincerity and utter love. My father was with me at the time, and the
presence of a spirit from so high a dimension left him bereft of
speech. When a high spirit makes an earthly appearance in this
way, it is within a numinous radiance that causes one's own
body to become very warm, and the words it speaks are so
filled with truth and light that one is moved to tears.

The following month, July, the secret repository of my
subconscious was unlocked, and the hidden part of my consciousness—Gautama Siddhartha, Shakyamuni—began to
talk to me in a mixture of Japanese and ancient Indian, urging
me to take my destiny upon myself and to spread the word of
Buddha. He revealed to me that I was an incarnation of El
Cantare, the focal consciousness of the Shakyamuni group,
and explained that it was my mission to be the salvation of all
living creatures through the worldwide revelation of the Truth.
The role of the Grand Nyorai Shakyamuni was twofold, he
said. There is the side represented by the Nyorai Amida (the
Savior), which consists of love, compassion, and faith. There
is also that represented by the Nyorai Mahavairocana (the
essence of Buddha), which is enlightenment, spiritual learn-

ing, and the secret knowledge of the spiritual domain. If the first aspect were to dominate within me, I would duly become a Grand Savior, but if the second aspect predominated I would instead become the Mahavairocana Buddha (the Great Enlightening), surpassing the Vairocana of the Garland or Mahavairocana Sutras.

I was utterly taken aback by it all. My upbringing had certainly been religious, and the existence of the spiritual realm I accepted as a proven fact. But this spiritual experience was so overpowering, and the scope of the mission set for me so enormous, that I could not conceal my shocked amazement. The only thing I understood at once was that I was a reincarnation of Buddha, and that it was to be up to me to reorganize the high spirits in heaven while also integrating all the various religions on Earth to create a new world religion. It was for me to gather all the peoples of the world into this new faith, to see to the development of a new civilization, and so herald the advent of a new age for the world.

I felt I was not yet ready. I needed time both to set about exploring the world of the spirit and to gain ordinary experience in my human life. I, therefore, decided to go on working in my current employment until I reached the age of thirty. It was then that, although I was experiencing no little inner turmoil, outward events in my life took a turn toward the mundane and the humdrum. For a year from 1982, I was dispatched to my company's headquarters in New York as a trainee—I, who had received a message from Jesus and been commissioned for a great undertaking by Buddha, spent many a day working with my fellow operatives on Wall Street in the steamy networks of international finance.

After a hundred hours of private English tuition at the Berlitz School of Languages I passed the entrance exam to the City University of New York, and went on to study international finance in a class of native English-speakers. I mingled with young businessmen and women in their early thirties from such companies as the Bank of America, Citibank, and Merrill Lynch, and together we studied all the ins and outs of the foreign currency exchange system. But I was less than happy in what I was doing. An unbridgeable gap was forming between the realities of my everyday life and the equally real events of my spiritual experiences. Sometimes, I would find myself looking up at the World Trade Center in Manhattan where I worked and wonder which was truly real: these huge buildings that seemed to touch the sky or the voices I could hear in my heart. My faith and even my sense of identity were severely tested.

The year I spent as a trainee was nonetheless a great success, and my superior asked me how I would respond if he was to request that I be posted there full-time. It was an unprecedented opportunity, and strongly implied that I was already on the way to the very top of the professional tree. But I was more interested in the manuscript I was then working on, a compilation of the spiritual messages that I had been receiving. So I turned the offer down, and recommended a junior colleague as an eminently suitable substitute—an act that was apprehended as one of extraordinary selflessness and generosity in the business world, but an act that for me represented a decisive step forward toward my destiny as a religious leader. I returned to Japan.

There I spent the next two years preparing myself for the task that lay ahead. In 1985, I published *The Spiritual Message of Nichiren*, a book that was soon followed by *The Spiritual Message of Kukai*, *The Spiritual Message of Christ*, *The Spiritual Message of Amaterasu-O-Mikami*, and *The Spiritual Message of Socrates*. I was still in the employment of the trading organization, so these books were published under my father's name. My own name appeared only as that of a contributory writer.

But the time finally arrived. In June 1986, Jesus Christ, Ame-no-Minakanushi-no-Kami, and other spirits came down to me one after the other and declared that now was the time to stand forth and announce my presence publicly. Then, on July 15 that year—just a week after my thirtieth birthday—I handed in my resignation to the company and took my first step on the wide plains of freedom.

Toward the end of the next month, August 1986, I started work on the first version of *The Laws of the Sun* (in Japanese) and completed it by the beginning of September. In October, I began to write *The Golden Laws*, and finished that also the next month. Publication of these works in which I proclaimed my message attracted quite a following of people who were in all sincerity searching for the truth.

12. Assembling a Congregation

The first lecture meeting I held was at the Ushigome Public Hall in Tokyo on March 8, 1987. Some 400 people turned up to hear me give a talk on "The Principles of Happiness," in which I introduced the four basic principles that constitute the core of what I want to teach—the principles of love, wisdom, self-reflection, and progress.

That month, I also planned the future development of the movement I had inaugurated. The first three years were to be dedicated to the study of spiritual law in its most basic form, the training of religious instructors, and the establishing of operational policies as an independent movement. Thereafter, we would concentrate on spreading the message as widely as possible, aiming for worldwide expansion.

April 1987 saw the publication of the first issue of a monthly magazine which contained elements of my lecture and some of my other writings, and which generally monitored the direction and progress of the movement. Further lecture meetings and study sessions were held concurrently, and bore fruit in gathering an increasingly committed membership.

As for my personal history, in spring 1988, I married my wife Kyoko, who had recently graduated in English Literature at the University of Tokyo, providing my life with a stable basis that enabled me to concentrate even more on my task. My wedding was blessed by the presence of a multitude of people and turned out also to be instrumental in the development of the organization. In due course, I discovered that the two of us had been life-partners in previous incarnations, notably in Atlantis, in the ancient Inca Empire and in ancient Greece.

Their hearts and minds moved by my impassioned lectures all over Japan, my audiences continued to expand in number. In 1988, there was not room for them all in Hibiya Public Hall, which has seating for 2,000. In 1989, the audience completely filled the Kokugikan in Ryogoku, which seats 8,500. In 1990, the exhibition hall at Makuhari, which holds well over 10,000, was filled to capacity every time I went to talk there.

Eventually, on 7 March 1991, four years after the initial lecture meeting, the Institute for Research in Human Happiness (IRH) received official recognition of its status as a religious organization. The movement I had inaugurated now had a new authority to go out into the world. The symbolic head of the movement is the Grand Nyorai Shakyamuni, also known as El Cantare, leader of the high spirits of the ninth dimension and reincarnation of the most revered Buddha.

The first annual public celebration of my birthday was held in the Tokyo Dome in July 1991, and was attended by no fewer than 50,000 believers. Such an attendance meant that within the very year it was officially incorporated as a religion, the Institute became one of the largest religious bodies in Japan, an expansion in membership so rapid that it has never before or since been equaled by any other religious grouping.

At the celebration I revealed to the congregation that I was in truth El Cantare, and described the mission I had been given as the Buddha of the Great Vehicle (Mahayana).

In September, the Revolution of Hope was inaugurated in order to sweep away the dark clouds that hang over the Japanese news media, and to cleanse the people of Japan of the spiritual pollution that persistently contaminates them. This represented a turning-point in the battle to create a Land of Buddha in post-war Japan.

In 1992 and 1993, I continued to affirm my teachings and their basis in Buddhist principles, and made use of satellite technology to give lectures via TV throughout the country simultaneously. During the same period, the Revolution of Hope attracted followers all over Japan, and the total number of believers approached ten million.

Our "Miracle Plan," in operation between 1991 and 1993, was also a tremendous success and contributed toward the Institute's becoming thoroughly accepted as a religion based on faith in El Cantare.

I now have to tell all the people in the world about the incarnation from the ninth dimension of El Cantare, and about His mission on Earth. It is the Coming, the Advent, the appearance on Earth of the highest Buddha, the supreme Savior. The world is in the process of undergoing purification, and through acceptance of El Cantare humanity can achieve the highest form of salvation.

Believe in El Cantare, and come gather round me.

Convey this message to all the people of the world. El Cantare is your Eternal Master.

Postscript

You have just read a book without parallel in the world, in that it gives a lucid explanation of the creation of the universe, the stages of love, the structure of enlightenment, and the rise and fall of past civilizations. Furthermore, it has revealed the true role of El Cantare. I advise you all to believe what is contained in this book, for one day in a future life, you will encounter it as Scripture.

Please bear in mind when trying to understand this book that when I use the term "Buddha," in preference to "God," it is because the expression is more in line with our fundamental teachings.

Having read this book, many of you will have resolved instantly what still remained as questions after reading any number of other volumes of religious and philosophical works.

I sincerely hope that this book will help people all over the world to awaken to Truth and to fulfill their mission of salvation. It will illuminate the light of truth in people's minds and the principle of hope for a future full of happiness.

At the time of writing, the Institute for Research in Human Happiness has now grown to consist of many more than ten million members all over the world. As the largest organization bringing enlightenment in Japan, it exerts leadership not only in the field of religion but also of politics and economy.

Ryuho Okawa
President
The Institute for Research in Human Happiness

ABOUT THE AUTHOR

Ryuho Okawa, founder of the Institute for Research in Human Happiness (IRH), Kofuku-no-Kagaku in Japanese, has devoted his life to the exploration of the Truth and ways to happiness.

He was born in 1956 in Tokushima, Japan. He graduated from the University of Tokyo. In March 1981, he received his higher calling and awakened to the hidden part of his consciousness, El Cantare. After working at a major Tokyo-based trading house and studying international finance at the Graduate Center of the City University of New York, he established the Institute in 1986.

Since then, he has been designing spiritual workshops for people from all walks of life, from teenagers to business executives. He is known for his wisdom, compassion and commitment to educating people to think and act in spiritual and religious ways.

He has published over 300 books, including *The Laws of the Sun, The Golden Laws, The Laws of Eternity, The Essence of Buddha,* and *The Starting Point of Happiness.* His books have sold millions of copies worldwide. He has also produced successful feature-length films (including animations) based on his works.

The members of the Institute follow the path he teaches, ministering to people who need help by sharing his teachings.

WHAT IS IRH?

The Institute for Research in Human Happiness (IRH), Kofuku-no-Kagaku in Japanese, is an organization of people who aim to cultivate their souls and deepen their love and wisdom through learning and practicing the teachings (the Truth) of Ryuho Okawa. The Institute spreads the light of Truth, with the aim of creating an ideal world on Earth.

The teachings of the Institute are based on the spirit of Buddhism and also embody a grand base for the integration of the major religions of the world, including Christianity. The two main pillars are the attainment of spiritual wisdom and the practice of "love that gives."

Members learn the Truth through books, lectures and seminars to acquire knowledge of a spiritual view of life and the world. They also practice meditation and self-reflection daily, based on the Truth they have learned. This is the way to develop a deeper understanding of life and build characters worthy of being leaders in society who can contribute to the development of the world.

SELF-DEVELOPMENT PROGRAMS

Video lectures, meditation seminars, study groups are available at local branches. By attending seminars, you will be able to:

- Know the purpose of life.
- Know the true meaning of love.
- Learn how to maintain peace of mind, instead of being swayed by anger or anxiety.

- Learn how to overcome life's challenges, such as difficult relationships with others, sickness, financial worries, etc.
- Learn to understand the workings of the soul and secrets of the mind.
- Learn the true meaning of meditation and its methods.
- Learn how to create a bright future within your family or at work.
- Know the Laws of success and prosperity.

And more…

IRH MONTHLY MESSAGES

This features lectures by Ryuho Okawa. Anyone is able to sub-scribe to the IRH Monthly Messages. Back issues are also available upon request.

MEDITATION RETREAT

Educational opportunities are provided for people who wish to seek the path of Truth. The Institute organizes meditation retreats in Japan for English speakers. You will be able to find keys to solve the problems in life and restore peace of mind.

For more information, please contact our branch offices or your local area contact.

Kofuku-no-Kagaku, USA
The Institute for Research in Human Happiness

New York
725 River Road, Suite 58
Edgewater, NJ 07020
Tel: 1-201-313-0127
Fax: 1-201-313-0120
Email: ny@irh-intl.org

Los Angeles
350 South Crenshaw Blvd.,
Suite A205
Torrance, CA 90503
Tel: 1-310-782-7776
Fax: 1-310-782-7775
Email: la@irh-intl.org

San Francisco
1299 Bayshore Hwy 200A
Burlingame, CA 94010
Tel / Fax: 1-650-347-7077
Email: sf@irh-intl.org

Hawaii
1259 South Beretania Street,
Suite 19
Honolulu, HI 96814
Tel: 1-808-591-9772
Fax: 1-808-591-9776
Email: hi@irh-intl.org

Chicago
Email:chicago@irh-intl.org

Boston
Email: boston@irh-intl.org

Florida
Email: florida@irh-intl.org

Albuquerque
Email: abq@irh-intl.org

KOFUKU-NO-KAGAKU
THE INSTITUTE FOR RESEARCH
IN HUMAN HAPPINESS

Tokyo
1-2-38 Higashi Gotanda
Shinagawa-ku
Tokyo 141-0022
Japan
Tel: 81-3-5793-1729
Fax: 81-3-5793-1739
Email: tokyo@irh-intl.org
www.irhpress.co.jp

London
Room T, 2nd Floor
Warwick House
181/183 Warwick Road
London W14 8PU
U.K.
Tel : 44-20-7244-6199
Fax: 44-20-7244-7648
Email: eu@irh-intl.org

Sydney
P.O.Box 437, Lane Cove
NSW 1595
Australia
Email: sydney@irh-intl.org

Melbourne
Email: mel@irh-intl.org

Toronto
484 Ravineview Way
Oakville, Ontario L6H 6S8
Canada
Tel: 1-905-257-3677
Fax: 1-905-257-2006
Email: toronto@irh-intl.org

Vancouver
Email: vancouver@irh-intl.org

Sao Paulo
(Ciencia da Felicidade do
Brasil)
Rua Gandavo,
363 Vila Mariana
Sao Paulo, CEP 04023-001,
Brazil
Tel: 55-11-5574-0054
Fax: 55-11-5574-8164
Email: sp@irh-intl.org

Seoul
3F, 98-52 Kalwol-dong
Yongsan-gu, Seoul
Korea
Tel: 82-2-762-1384
Fax: 82-2-762-4438
Email: korea@irh-intl.org

Taiwan
5F, No.109-7
Hsin yi Road, Section 3
Taipei
Taiwan
Tel: 886-2-2705-8097
Fax: 886-2-2705-8302
Email: taiwan@irh-intl.org

Hong Kong
Email: hongkong@irh-intl.org

Other Asian Countries
Email: asia@irh-intl.org

Lantern Books by Ryuho Okawa

The Golden Laws
History through the Eyes of the Eternal Buddha
1-930051-61-1
Lantern Books, 2002

The Laws of Eternity
Unfolding the Secrets of the
Multidimensional Universe
1-930051-63-8
Lantern Books, 2001

The Laws of Happiness
The Four Principles for a Successful Life
1-59056-073-6
Lantern Books, 2004

The Starting Point of Happiness
A Practical and Intuitive Guide to Discovering Love,
Wisdom, and Faith
1-930051-18-2
Lantern Books, 2001

The Origin of Love
On the Beauty of Compassion
1-59056-052-3
Lantern Books, 2003

Love, Nurture, and Forgive
A Handbook to Add a New Richness to Your Life
1-930051-78-6
Lantern Books, 2002

Invincible Thinking
There Is No Such Thing As Defeat
1-59056-051-5
Lantern Books, 2003

An Unshakable Mind
How to Overcome Life's Difficulties
1-930051-77-8
Lantern Books, 2003

Guideposts to Happiness
Prescriptions for a Wonderful Life
1-59056-056-6
Lantern Books, 2004

Tips to Find Happiness
Creating a Harmonious Home for Your Spouse,
Your Children, and Yourself
1-59056-080-9
Lantern Books, 2004

Want to know more?

Thank you for choosing this book. If you would like to receive further information about titles by Ryuho Okawa, please send the following information either by fax, post or e-mail to your nearest IRH Branch.

1. Title Purchased

_____

2. Please let us know your impression of this book.

_____
_____

3. Are you interested in receiving a catalog of Ryuho Okawa's books?

       Yes ❑   No ❑

4. Are you interested in receiving IRH Monthly?

       Yes ❑   No ❑

Name : Mr / Mrs / Ms / Miss : _____

Address : _____
_____
_____

Phone: _____

Email: _____

Thank you for your interest in Lantern Books.